# Data Communications Network Auditing

## How to Save Hundreds of Thousands of Dollars and Still Improve

## Network Response Time Network Performance and Network Availability

by the man who's done it several times before
Bruce Griffis

1

CRC Press
Taylor & Francis Group
Boca Raton London New York

CRC Press is an imprint of the
Taylor & Francis Group, an **informa** business

CRC Press
Taylor & Francis Group
6000 Broken Sound Parkway NW, Suite 300
Boca Raton, FL 33487-2742

First issued in hardback 2017

© 1996 Bruce Griffis
CRC Press is an imprint of Taylor & Francis Group, an Informa business

No claim to original U.S. Government works

ISBN 13: 978-1-138-41217-0 (hbk)
ISBN 13: 978-0-936648-93-4 (pbk)

**Visit the Taylor & Francis Web site at**
**http://www.taylorandfrancis.com**

**and the CRC Press Web site at**
**http://www.crcpress.com**

# Foreword

## by Harry Newton

Bruce Griffis worked his datacom auditing magic on one company and saved them over $300,000 a year. What was more remarkable than the money was that after his magic the network got better in all respects.

I asked Bruce to write this book because we all need a little Bruce Griffis Magic in our networks.

Saving money and improving our networks will make us heros before our management. It will make our customers love us more. What more can you ask?

The Bruce Griffis Magic is not rocket science. Doing it is like completing a gigantic crossword or playing chess. It's rigorous. It's intellectual. It's fun.

Read. Enjoy. Profit.

*Harry Newton*

Harry Newton
October 1996
New York, NY

# Introduction
by Bruce Griffis

The title of this book should bring up a few questions for you. Can you reduce your network communications bill? Can you really save hundreds of thousands of dollars? Will it be an arduous task? Do you have to be a network engineer?

The answers are straight-forward. If you enjoy puzzles, if you truly like the "parts and pieces" that make up a network, and if arcane bills, invoices and EDI reports don't throw you off - you will do well!

I followed the procedures in this book and found substantial savings whenever and wherever I applied them. If you have a Wide Area Network with multiple network components - you will be surprised at how quickly the savings add up!

How will you reduce bills? By finding circuits no longer in use. By finding offices with excess capacity. By finding disconnects that never occurred. By finding billing errors. By merging single protocol networks into a multi-protocol network.

How will you improve network response time? By optimizing your network after you determine what you have and comparing it with what you need. You will now find money in your budget that was never available before!

How will you improve network performance? By knowing EXACTLY what connects to what, where it connects, and what it costs - you can determine where your company's money should be spent. You could reduce speeds in small offices, remove unneccessary circuits, and use the savings to increase circuit speeds where required.

How can you improve network availability? By knowing ALL relevant circuit IDs, router ports, frame relay DLCIs and network connectivity - you can reduce network problem determination times.

A good network audit can help you reduce your company's network expenses. This savings goes right through to the bottom line! Read the book. Follow the procedures. Improve your company's financial performance by reducing your network's impact on the bottom line.

A network is a living, breathing organism. It expands and contracts to meet your needs. It carries the lifeblood of your company -- information. Your job is to make sure that there is an optimum pathway for that information.

That pathway should be enough to meet your needs, but never more than you need. It must fit within your company's budget, but never take more than is absolutely necessary.

A network audit is a perfect point to build a case for better network management tools. You will have a cost saving. That savings may be able to finance that network management station you have always wanted. With your new network management station, you will be able to start trending and reporting on critical network statistics: availability, reliability, performance, average utilization. You will be able to plan for higher bandwidth where required, and for less bandwidth as personnel shift from one office to another. You can only win!

# Author's Notes

This book is designed to help you understand the "parts and pieces" of communications. This book is designed to help you determine how components fit together, and what they look like on your bill.

You should be able to look at your invoices and know exactly what each item is. You should also know where it fits in your network. If you have a circuit bill, it should relate to two networking components - a component on either end. If your bill breaks out port charges, access charges, CSU charges and PVC charges - you should be able to relate every billed item to a network device. That network device should be up, operational, and passing data.

As you determine what network devices you have, you should get an idea of what clients reside on your network, and what application servers reside on your network. You will also gain a clear idea of what each networking component costs you each month.

This information is critical. You may find small offices that have more networking "horsepower" than is required. You may find larger offices that just don't have a fast enough network. You may be able to determine the cost of getting to an interactive application - and determine that the value of that information just isn't worth the cost of the network! Conversely - you will be able to determine how much it will cost to increase circuit speed and improve performance - providing you are communicating with well-behaved wide area network aware applications!

By knowing what you pay to network each office, you can determine the average cost of providing network connectivity. You can also determine the $90^{th}$ percentile. This will let you estimate, within a 90% range of accuracy - what it will cost to bring the next office up on your wide area network. That way you will not look bad when you are surprised at your next Information Systems meeting. Your CIO may (and probably will) ask you what it will cost to add East Podunk to the network. You will be able to state what your router will cost, and estimate within a 90% degree of accuracy what your wide area network costs will be. You should also have a clear idea of what the intervals are to provide network connectivity. You will be able to say that the router will cost $2,700.00, the circuit will cost $780.00 a month, and that it will take 45 days to get it up and running.

# How to use this book

This book can be used a general guide to data communications auditing techniques. This book will show you how to determine if a circuit that appears on you bill is actually installed, connected to network equipment, and in use.

This book can be used as a guide in establishing a data communications auditing business. The proposal at the end of this book can be used as a guide in creating your own proposals. The forms included in this book can be freely copied and used as you audit a network.

This book deals with physical and logical network connectivity:
  what it is
  how devices connect
  how you determine if they are passing information

This book also contains product specific information based on Cisco router command line interface, and IBM's Netview.

While this book can be used by itself, your  network auditing and optimization efforts will be greatly enhanced by referring to other resources as noted in the book.

# Who Should Read This Book

This book is intended for Network Managers and Telecommunications Professionals new to the network validation and budgeting process. This book is also intended for Telecommunications auditing professionals who want to add data communications auditing to their repertoire.

Help Desk and Network Control Center personnel will find this book useful. It will help you to tie physical network elements with logical network elements, such as circuit IDs and billing detail. Nothing feels better than calling a vendor with a problem and knowing exactly what you are paying that vendor. I keep a listing of router ports and attached circuits and prices - so I know financial relationships while escalating network outages.

Finally, network billing analysts should find this book helpful. You may find that you are not properly billing all of your customers for network connectivity. It happens more often than one would expect!

This book does not assume that you have a great deal of knowledge of the day-to-day tasks involved in verifying network connectivity and relating it to network bills. This book includes samples where applicable. IBM SNA samples are from IBM's Netview application. Multi-protocol samples are from the command line interface of Cisco routers. By learning what you are looking for, you will be able to determine how to look for it on your communications network and with your hardware and software mix.

If you are a Telecommunications auditing professional, you may want to invest in manuals or CDs from Cisco and Wellfleet. Alternatively, you may want to just monitor the Cisco and Wellfleet news groups and ask very basic questions. Just don't advertise there! This should let you serve a wide range of clientele. You do not need to take courses in networking to perform a wide area network audit, although some elementary training would be beneficial.

You will need to know how to determine the status of serial interfaces on routers. You will also need to know how to determine the status of IBM SNA/SDLC devices if your client has a legacy network. You will find other wide area networking equipment, but should be able to handle differences in command interfaces once you know what you are looking for. Other than that, you're good to go!

# Table of Contents

# Part I - Getting Started

## - Introduction

**What is a Data Communications Network Audit**
A Data Communications audit is used to determine what network hardware you have in place, how it connects, and how it appears on your bill.

While performing a data communications audit, you will learn a lot about the wide area network. You will become extremely familiar with a company's choice in communications equipment. If you have only been involved with software, you will learn about the hardware components. If you are only familiar with hardware, you will have the opportunity to learn some basic software functions. If you have never been involved in the financial aspects of a network, you will learn a whole new area!

This book is primarily concerned with the wide area network. There are several opportunities in Local Area Networks as well. As Wide Area Networks are connected via circuits which have monthly re-occurring charges, Wide Area Networks provide a wealth of opportunity to the network auditor.

Auditing and optimizing a wide area network is like fitting a puzzle together. The pieces of the puzzle include:

What are the application servers?
Where do the application servers reside?
What applications reside on the servers?

What are the clients?
Where do the clients reside?
How do clients communicate with the servers?
What protocols are used between the client and the server?

Who should be contacted if an end-node goes down?
Who is the network carrier?
What are the circuit IDs?
Who do you contact at the remote office?

What are the connecting pieces?
Are the components routers
                    bridges
                    gateways
                    repeaters
                    or simple hubs?

What is the client environment?
                    Is it Ethernet?
                    Is it Token Ring?
                    Is it Fast Ethernet? ATM? FDDI?

What is the cabling?
                    Is it CAT5?
                    Is it Level 3?
                    Is it Type 1
                    Is it ThinNet?
                    Is it ThickNet?

How do end nodes communicate over a distance?
                    Is Frame Relay employed?
                    Are leased lines used?
                    Is ISDN used?
                    Is Switched 56 used?

What are the speeds?
What percentage of capacity is being used?
What will it cost to increase speeds?
What is the savings with reducing speeds?
Do you need to increase speeds?
Can you reduce speeds?
Can circuits be disconnected?
Can circuits be multiplexes on a T1?

Can networks be merged?
If I have routers supporting TCP/IP and IPX/SPX, can I transport
SNA/SDLC traffic? If so, what is the impact on the network?
What is the impact on the budget if networks are merged?

## The Relationship with Voice Network Auditing

You may want to perform a voice communications audit while performing a data communications audit. There are good books on billing practices and traffic engineering.

While performing your voice audit, you can start to look for excess capacity on voice T1 circuits. The excess channels can be multiplexed off using a drop-and-insert mux to support your data network. This allows you to reduce your T1 access costs.

Voice networks should not be engineered based on billing data only. It is critical to dial into a PBX and determine how trunk groups are arranged. You also need to look at Least Cost Routing plans to determine how calls are routed.

If you are using T1s, you will need to know if your 800 numbers are sent in over dedicated channels, or if they are non-channelized and routed using DNIS. After clearly knowing your voice and data networks you can start to merge them.

Data communications auditing bears similarities with voice auditing. The questions are of the same type, simply asked in a different jargon. The questions are:

> Can all the circuits on the Customer Service Record be identified?
> Are they all in use?
> Can any be combined?
> If you "call" them, do they "ring" at the customer's premises?

> By calling, you send a packet of data using TCP/IP, IPX/SPX, AppleTalk, ...

> The customer's premises in this case would be a server - be it a large server like an IBM 3090 using VTAM and SNA protocols, or a server like Novell's NetWare or Microsoft's LAN Manager or NT.

Voice T1s are identical to data T1s. The circuit is the same. The signaling can be the same (D4/AMI, ESF/B8ZS). The CSU/DSU can be the same. The connecting devices are different - a PBX versus a router or bridge, and the connecting cable may be slightly different. The difference is in the use of the circuit, and the measurements of that usage.

### The Relationship With Video Auditing

Video networking is a primary application of ISDN. You will use either 2x64 (ISDN BRI) or H0 (384K, or 6 channels of a PRI) for video conferencing in most instances.

If you are using ISDN PRI with H0 formats, you will have spare capacity on your ISDN T1 for other applications. Your audit will show you where you could multiplex out excess channels of your PRI for an ISDN BRI pool, or pick up a few channels of clear channel for more traditional data networking.

### The Audit as a first step in network optimization

While performing your network audit, you will also be able to start to optimize the network. This includes reducing access speeds where offices have been reduced in size, and increasing access speeds where networking requirements have grown.

Changing communications methods from leased lines to frame relay, or to ISDN BRI is part of a network re-design. Re-designing the network brings substantial cost savings potentials, and can greatly improve performance.

This book deals with network auditing and optimization. This book does not serve as a network design manual, but does give basic pointers where applicable.

Offices expand and contract as business needs dictate. The network needs the same flexibility. If you plan to optimize the data communications network, you should consider optimizing the voice communications network as well.

If your reports point out an office that is seeing reduced network utilization, it would be beneficial to review voice trunk usage as both voice and data trends tend to correlate.

There are several tools available to help optimize a wide area network. These tools run on either Unix or Windows operating systems. There is a wide range in costs for these tools.

Some are available as freeware, using perl scripts and running on a Unix platform. Some are available at a nominal cost (appr. $5K for Stony Brook Software's Wide Area Network Manager), and some are expensive. The tool set you chose depends on your specific requirements and budget, and on the operating system you are most comfortable with.

Several frame relay vendors provide statistics on network utilization. These are good tools for optimizing a frame relay network. As these reports are generally available at either no cost or a nominal fee, they should be used.

Some vendors average utilization over a wide range in time. As data communications is often spikey in nature, reports averaging large time spans can misrepresent requirements.

For example, MCI's reports track utilization and throughput during the busiest hour of the day, using a bouncing busy hour. These reports are excellent. Other reports average utilization over a 4 hour time frame. This large a time span can be meaningless.

### Approaches to Data Communications Auditing
There are several approaches you can take towards performing a network audit. The approach you take depends on relationships. Some of these relationships are: project leader or internal company consultant, or as a consultant hired for this specific project. You could also be a professional telecommunications auditor, expanding your business into data communications.

### Project leader or Internal consultant
As a project leader within the company, you may already have built relationships with people on the help desk, in the Network Control Center, and in remote offices. You may have a relationship with senior management. Performing a network audit will help you directly effect your company's bottom line.

You will be in a good position to perform an audit, and you may find it relatively easy to gain access to network components you don't normally have access to. Performing a network audit would be a great first step towards becoming more involved with the network.

### The - External Consultant
As an external consultant, you will need to market your services to a company.

You will need to discuss how thorough network documentation can reduce problem resolution time. You will want to discuss the importance of auditing network bills, as monthly re-occurring circuit charges add up very quickly. You will need access to components that are normally secured from outside consultants.

Some of the components you will need access to include:
> The companies Information Management system
> Netview, if the company uses IBM SNA devices
> Browse access to routers
> SNMP get access

---

You will also need physical access to a company's computer room, and telecommunications closets.

As you approach the project, you will need to keep in mind that the people that hire you may be the people involved in creating and managing the network. Discretion is always the key word. You may want to stress that documenting and auditing the network can be done by a consultant. This frees the network staff to support the network and can also make it easier for the networking staff to perform their job.

## The Telecommunications Auditor
As a professional auditor, you may currently bill clients based on a portion of your savings. This can be a good approach.

An alternative is to look at network auditing as a project consisting of three phases. You may wish to be compensated in a different manner for each phase.

You could also bid on each phase separately, allowing the client to determine how much of the project they will do, and how much of the project you will do. It can give the client an opportunity to reduce consulting costs, and give you an opportunity to maximize short-term earnings potentials.

Phase I would be the network audit. You could bill the client for a portion of any costs you recover. You could also bill the client for a portion of the first years savings. This can be significant in larger networks, in dynamically evolving networks, and in networks seeing reduced usage due to reduced business opportunities.

Phase II would be network documentation. As good documentation can help reduce the amount of time an office is down due to a network outage, you are performing a job that will reduce a company's exposure to losses due to the inability to get to information.

This may not have a hard dollar value. You should consider approaching this phase of the project as a consultant and estimate the amount of time required to update an information management system.

The skill set required for network documentation is different than the skill sets required for network auditing and network optimization. If you have a small partnership, you could use the network documentation position to train new people.

The third phase of the project would include network optimization. The benefit in reducing network circuit speeds in smaller offices is reduced network costs. The benefit in increasing circuit speeds where warranted is increased productivity. It is difficult at best to put a dollar figure on increased productivity. This phase should be approached as a consultant.

# - The Plan and Proposal

**The Business Plan**

An essential step in creating a service in data communications auditing is to develop a business plan. There are several good books on creating business plans. This is to serve as a simple introduction to the world of planning.

Any business venture needs to have a plan in order to succeed. The plan helps you define where you want to go. Your business plan should contain the following elements:

1. Executive Summary
2. Mission and Vision Statements
3. Strategies
4. Personnel
5. Objectives
6. Financial Plan

The Executive Summary is a brief, high level look at your business. It may be the only item that is read, so the summary should be able to get and keep the reader's attention.

Your mission statement describes what it is that you do. It is brief, and should be something that you can commit to memory. Along with a mission statement, you should have a Vision statement.

A mission is your purpose. In this case your mission is to audit the data communications network and identify cost savings. Your mission is also to document the network to facilitate network problem determination and lead to increased network availability.

Your Vision is an image of what it will be like when your mission is completed. Your vision may be to provide the best possible data communications network that meets your company's financial and network performance requirements.

Strategies are what you use to meet your objectives. This includes how you will audit the network, and whether you will use commercially available documentation packages or develop software on your own. This also includes your marketing strategy: how you will let the world know about your services.

When you identify personnel requirements, you will make decisions on whether to handle the project yourself, or to use assistance. This may be people within your organization, consultants, or sub-contractors. You should identify personnel issues before accepting that job to audit a network consisting of 300 routers and 28 frame relay networks supplied by 4 different vendors!

With the increasing popularity of the Internet, and with newsgroups dedicated specifically to telecommunications, network management and data communications - the concept of a virtual organization is not far from reality. By making careful inquiries and posting e-mail messages to specific newsgroup contributors, you will be able to find consultants and sub-contractors that can help you provide a more complete offering.

If you are strong in data communications, but do not have strength in tariffs, you should be able to find someone to complement your strengths. Alternatively, if you are good in tariff analyses, but not quite as strong in data communications, you should be able to find someone with the appropriate skill sets.

If you are auditing a national or even international company, you could create a virtual organization that can complete the audit, perform network optimization, and analyze tariffs. Once the job is complete, your virtual organization would then dissolve - only to come back together again when anyone in your group obtains another client.

Objectives are clear cut statements of what you intend to do. This is your working plan. Your objectives may be to: meet with senior management, meet with networking personnel, review network documentation, verify bills, and eliminate excess circuits. Your objectives should be used in developing your project plan.

The world of telecommunications management and optimization is a great candidate for using the tools of telecommunications in creating a virtual organization. Some of these tools include:

Audio conferences for your staff meetings
Internet E-Mail for communications
Internet FTP, or E-Mail attachments for documents
Internet chat or web phones for voice calls
CU-SeeMe for videoconferencing
Web pages to advertise your services

Your financial plan should include your estimated costs and income. It should be detailed. Points of your financial plan should correlate with points of your objectives.

Some of your costs may be:
Network utilization tool kit consisting of a portable or laptop, TCP/IP protocol stack
Possibly IPX/SPX protocol stack
Network trending software
Modem for dial-in support
PCAnywhere or Carbon Copy for out-of-band management
(this tool kit should be low end, as you would leave it on a
        customers premises for 30-45 days)

Desktop or laptop PC for reporting and trending
Application suite for creating documents, linking spreadsheets, creating graphic presentations.

Phone, fax, Internet account


## Internal Business Plan

Do you need a business plan if you are auditing your own network? Absolutely! An audit takes resources. This is time (which is money) and money (which is also money!). You need to identify what you intend to do, how you intend to do it, what it will cost, and what you expect to gain. By presenting your plan to management, you can gain approval to move ahead with the project. You can then use the plan to help drive your project.

You can then measure the success of your project, by seeing how closely it correlates with your plan, and by seeing cost savings.

## Consultants Business Plan

A business plan written from a consultant's viewpoint will need to be much more detailed. You will be trying to sell a service, and will need to know the value of your services, your objectives, and an approximation of your level of effort before making a presentation.


## The Proposal

A good proposal is a critical element in securing business. If you are performing network auditing as an entrepreneurial activity, it is highly recommended that you review a few books on writing competitive proposals. Alternatively, you can use the sample proposal at the back of this book!

If a potential client issues a request for proposal, it will be used to narrow the field of competitors. The potential client will look at how well you respond to his RFP and how well you address his concerns. The potential client will also look at your perceived technical competence, your project management skills, and the costs and benefits associated with your solution. Your proposal will need to set you apart from your competition.

Your response to the RFP should include a statement of work. This is a high level abstract that is used to show that you understand the client's requirements. Your proposal should also include a detailed description of what you will do and how it will benefit the client.

The proposal should include a brief project plan. The project plan should not include an exact description of all the work to be done, and how it will be done - it should cover the highlights.

Finally, your proposal needs to show why the client should select you. This includes a statement of your qualifications and competence in network auditing.

Your proposal should include some or all of the following items:

1. Cover letter
2. Title page
3. Table of contents
4. Executive summary
5. Body of Proposal
   5A. Technical detail
   5B. Deliverables
   5C. Cost justification and return on investment (ROI)
   5D. Schedules
   5E. Documentation
   5F. Resumes
6. Glossary
7. Appendix

A sample proposal can be found in Appendix C. You may modify it to meet your needs.

## Marketing
Successfully marketing a product such as data communications network auditing is as important as performing the job. Without a good marketing plan, there would be no auditing to be performed.

The personality traits, verbal communication skills and persuasiveness required for marketing a product a very different than the traits and skill sets required to perform a detailed audit and optimization projects.

That being said, is there anything you can do? You could meet with other telecommunications consultants. You could present your services to a headhunter. You could look for other telecommunications auditors to form a network. You could unobtrusively look for prospects in Internet news groups. You could advertise on a Web page. You could contact CIOs and CTOs of local companies.

# Part II

# An Introduction To Data Communications

## Legacy Networks

Legacy networks are networks that were built to support the hierarchical computing model. The most dominant legacy network is IBM's Systems Network Architecture. IBM SNA is different from Local Area Networks, in that there is typically one or more mainframe centers, and remote devices communicate to the mainframes in a tightly controlled environment.

SNA components include:
>    The mainframe computer, running VTAM
>    IBM 37x5 Front End Processors running ACF/NCP, and possibly
>        NTRI and NPSI
>    IBM 3x74 Control Units
>    3x78 and 3x79 terminals
>    Printers

SNA components can also include the IBM AS400 midrange computer and its networking components.

Terminals are connected to control units using either coax or twisted pair and baluns. Controllers can concentrate traffic for up to 32 terminals. These terminals can have multi-session ability, allowing access to TSO on one session, CICS on another session, and IDMS on a third session. This access can be simultaneous.

Control units can reside on point-to-point circuits, or on multi-drop circuits. Control units can also attach directly to 37x5 Front End processors through modem eliminators. Another alternative is token ring attached control units. Rings can be local, by being served through a TIC (Token Ring Interface Coupler) on a 3745. Rings can also be remote, and be connected via remote bridges.

IBM 3x74 control units can be channel attached to IBM mainframe computers. This is normal in computer centers. Channel attachment is used for local controllers, it is not normally used for remote controllers, although a channel extender and a T1 can be used to connect a "local" controller in a remote location.

Terminals only communicate with control units. Control units communicate with front end processors. Front end processors communicate with mainframes. This model still serves well for high volume transaction processing applications.

There are still several large networks of IBM SNA devices, as the equipment and the protocol serves a specific purpose very well. There will be SNA networks for some time to come, although there will be more integration of multi-protocol networks.

# IBM SNA/SDLC Environment

## IBM 3x74 Controllers serving 3x78 / 3x79 terminals and printers

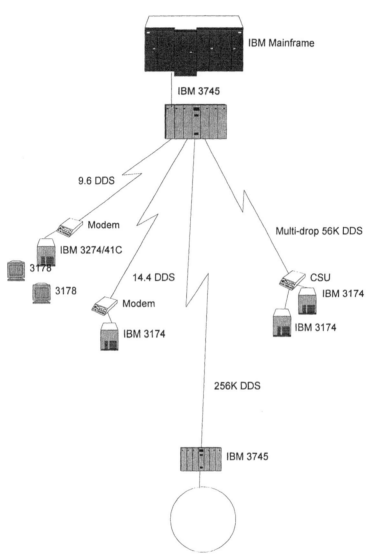

Remote bridge serving 3174/13R controllers and Token Ring PC's w/ PC3270

## The basic components of a LAN

As PC use has grown, so has the need to allow Personal Computers to communicate with each other. PCs communicate with other PCs in the same building over a Local Area Network (LAN). As PCs communicate with geographically disbursed PCs, they communicate over a Wide Area Network (WAN) or a Metropolitan Area Network (MAN).

A Local Area Network consists of several components. Each component is designed to serve a specific purpose. LANs typically use either Ethernet or Token Ring as an access method. There are other access methods, including ARCNet, AppleTalk, FDDI and ATM. Ethernet also comes in a number of varieties, including 10Base-T, 10Base-2, 10Base-5, Switched Ethernet and Fast Ethernet.

## Ethernet

Ethernet follows the IEEE 802.3 standard. It uses an access method called CSMA/CD, or Carrier Sense Multiple Access with Collision Detection. Under the CSMA/CD access method, when a PC wants to send or request information over the LAN, it listens for traffic on the LAN. If it does not sense a packet on the LAN, it is free to transmit. If another device is transmitting at the same time, a back-off algorithm is used to determine when to retransmit. Due to the nature of CSMA/CD, Ethernet segments tend to be relatively small.

Several manufacturers have addressed some of the issues concerning Ethernet. This includes Switched Ethernet and Fast Ethernet. With a switching hub, you can configure a port to operate as either half duplex or full duplex. You can also dedicate up to a full 10 MBPS to the port. This helps to reduce collisions.

Fast Ethernet is another alternative. Using Fast Ethernet, you can operate a LAN device at up to 100 MBPS. By using a switching hub that supports Fast Ethernet, you can dedicate 100 MBPS to a single device on the LAN. This is used to support heavily utilized servers or imaging devices.

Fast Ethernet is also used in building backbone Ethernet segments between floors in a building, and in connecting segments in a campus environment.

## Token Ring

Token Ring is another topology used for Local Area Networks. Token Ring was originally developed by IBM, and is found predominantly in IBM shops. Token Ring uses the IEEE 802.5 standard, which supports token passing. A token is sent out on the LAN. Each device on the LAN checks to see if the token is empty. If the token is empty, the device can use the token to transmit information.

Token Ring can operate at high utilization levels, and at higher data rates than 10Base-T Ethernet. Token Ring is also normally more expensive. Token Ring NIC cards typically cost more than 10Base-T Ethernet NIC cards. Token Ring hubs typically cost more than 10Base-T Ethernet hubs. Token Ring's main saving grace - speed and high loading abilities, are giving way to switched Ethernet and Fast Ethernet. Because of this, Ethernet is making in-roads in more traditional Legacy environments.

Token Ring still serves a purpose, and will be found in computer centers that use IBM mainframes, and/or IBM AS400s.


## FDDI

Fiber Distributed Data Interface is a standard used to build 100 MBPS fiber optic local area networks. FDDI uses an enhanced token passing protocol, similar to 802.5 Token Ring. As FDDI uses a token passing mechanism, rather than a CSMA/CD standard, FDDI can support high loading levels.

FDDI is built using a dual counter-rotating ring design. FDDI serves well as a backbone for connecting routers and servers. FDDI is not well suited to supporting user segments of Local Area Networks due to the cost of fiber, and the cost of appropriate NICS.


## Wide Area Network Components

As devices on different LANs communicate with each other, they do so over components such as repeaters, bridges, routers and gateways. These devices operate at different layers of the OSI model. The OSI model uses 7 layers to define how different devices inter-operate. The layers follow.

## OSI Model

| 7 | Application |
|---|---|
| 6 | Presentation |
| 5 | Session |
| 4 | Transport |
| 3 | Network |
| 2 | Data Link |
| 1 | Physical |

Layer 1 is the physical layer. Repeaters are devices that operate at the physical layer. A repeater is used to extend a LAN, but does no filtering or routing. Bridges operate at layer 2, while routers operate at layer 3. Gateways operate across the range of layers.

**Bridges**

Bridges operate at Layer 2 of the OSI model. A bridge can connect 2 different LAN segments together. The bridge can connect segments or rings that are in the same building (a local bridge), or LANs that are in different locations (remote bridges).

Bridges are generally simpler to configure and maintain than routers, and also tend to be less expensive. Bridges have some drawbacks - including broadcast storms, and limited abilities to isolate sub-networks.

Bridges are used to segment LANs and to filter traffic. This helps to improve performance of the Local Area Network by reducing traffic on the segment or ring. Learning bridges can automatically learn about the network by examining the packets they receive.

There are both Ethernet and Token Ring bridges. Token Ring bridges can be used for Local Area Networks, or for IBM SNA devices that support Token Ring and LLC2 protocols. This includes IBM 3174 control units such as the 3174/13R, IBM AS/400 midrange computers, and IBM 3745 Front End Processors with TIC cards.

A local bridge is built with a PC, 2 NIC cards, and software. One NIC card connects to segment A. The second NIC card connects to segment B. The bridge software performs filtering in determining what packets to forward, and how many hops to allow.

A remote bridge is built using 2 PCs. A PC in building A would have a NIC card connecting it to the local area network and a card connecting it to the wide area network. A similar PC would be in building B. The 2 PCs would be connected using a 56K, Fractional T1, or frame relay circuit.

There are also hardware specific bridges, that do not operate on PCs. These devices are purpose built. Routers can also act as bridges by using bridging software instead of routing software.

A bridge is used to connect an Ethernet LAN to an Ethernet LAN (although translational bridging can be used to connect dissimilar environments to varying degrees of success).

**Routers**
Routers are used to interconnect multi-protocol networks. Routers can support multiple protocols such as TCP/IP, IPX/SPX, and AppleTalk. Routers can also support multiple access methods for Local Area Network connectivity, including Ethernet, Fast Ethernet, Token Ring, FDDI and ATM. Routers communicate remotely using leased lines, Frame Relay, ATM or ISDN.

Routers communicate with each other through routing protocols. Some protocols operate across multiple vendors. These include RIP (Routing Information Protocol), OSPF (Open Shortest Path First), EGP and BGP (Border Gateway Protocol). Some proprietary protocols include Cisco's IGRP (Interior Gateway Routing Protocol) and EIGRP.

Routers provide more isolation between networks than bridges provide. Network isolation, segmentation and security can be done through access lists, network filters and firewall routers.

**Modem Pools**
Modem Pools are a basic component of data communications networking. A LAN may have a communications server to allow clients to dial out to other networks, such as online services. The communications server may also be used as a dial-in platform, allowing remote users to dial in to the local area network.

As a dial-out platform, the communications server provides each client on the network with a communications port. The ports all reside on the communications server. This saves the business money, as not all clients need their own analog line or modem.

As a dial-in platform, the communications server can provide e-mail access, or can provide access to applications residing on the LAN. Communications servers are typically protocol specific, and can sometimes be application specific.

A remote client may have to dial in to one server for e-mail, then dial into another server for applications. This can be reduced by creating a communications server that serves all the applications that reside on the LAN.

The opportunity for the network auditor is to determine the load on the communications server. By doing this, you can determine how many analog ports are required to serve the community.

Following is a sample from a modem pool used for e-mail access. This e-mail pool is served by an 800 number. The 800 number reduces costs to the business, as it has a lower cost per minute than using calling cards for modem access.

This specific modem pool has an 800 number through MCI, so we can use MCI's TrafficView to get an idea of the traffic levels and grade of service. Other vendors offer tools for measuring 800 number performance. If you have a modem pool served by an 800 number, it will be beneficial to find out what tools your specific carrier has.

## Inbound E-Mail Server Traffic Loading

Summary Report for 800 Numbers

| Date Range | Attempts | Comp | Incomp | Avg Dur |
|---|---|---|---|---|
| 06/01/96 - 06/30/96 | 2,616 | 2,215 | 401 | 1.3 |
| 07/01/96 - 07/31/96 | 2,388 | 2,146 | 242 | 1.4 |
| 08/01/96 - 08/28/96 | 2,385 | 2,202 | 183 | 1.4 |
| TOTAL | 7,389 | 6,563 | 826 | 1.4 |

Grade of Service Report

| Date Range | Grade of Service | Total Minutes |
|---|---|---|
| 06/01/96 - 06/30/96 | 15.3 | 2,846.10 |
| 07/01/96 - 07/31/96 | 10.1 | 3,071.57 |
| 08/01/96 - 08/28/96 | 7.7 | 3,144.15 |
| TOTAL | 11.2 | 9,061.82 |

From these reports the network auditor can learn the average length of time spent for e-mail access. At 1.4 minutes, we can see that the process is highly automated! We can also see that even though the number of minutes has increased each month, we have also improved the grade of service.

In determining how the calls should be handled, we have several items to consider. This includes:

Cost Per Minute (CPM) for calling from a small location where volume discounts may not apply.

Cost Per Minute for using a calling card. Also include any surcharges for using a calling card.

Cost Per Minute for dialing into an 800 number. You may get a lower rate if your headquarters office has a higher traffic volume.

By looking at the amount of traffic carried by the modem pool or terminal server, and looking at our calling options - we can determine the best way clients can call in.

By assuming that clients are calling from a small office using switched services, we can estimate that they are paying around 12 cents per minute for long distance. Another alternative is for remote users to call in using a calling card. There is currently a Brand X card on the market that runs at 17.5 cents per minute with no surcharge.

A third alternative is that the modem pool is located in a large office that has dedicated access to a secondary carrier. It would be safe to estimate that calls to their 800 number run around 8 cents per minute.

By looking at our minutes of traffic, we can see that it would cost the following to get to the modem pool:

| Access Type | June | July | August | Total | Annualized |
|---|---|---|---|---|---|
| Switched | 341.53 | 368.59 | 377.30 | 1,087.42 | 4,349.68 |
| Calling Card | 498.07 | 537.52 | 550.23 | 1,585.82 | 6,343.28 |
| Dedicated | 227.69 | 245.73 | 251.53 | 724.95 | 2,899.80 |

The savings by using an 800 number versus switched access amounts to $1,449.88. The difference between supplying e-mail users or remote access users with calling cards versus using an 800 number totals $3,443.48. Not bad for taking a look at a small dial-in system. As most business have more than one dial-in server, the savings potential is even greater.

From here we can determine the number of modems and analog lines required to serve the end-user community. This is determined by finding out the desired level of blocking (modem pools generally run in the 5 to 10% range), and using a traffic engineering application. We also need to calculate traffic loading during the busy hour.

In an e-mail or a voice mail example, calls are not 100% random. Callers tend to check e-mail or voice mail in the morning, after lunch, and again in the afternoon.

A simple Erlang B calculation assuming random arrivals tells us we need 2-3 modems to serve the traffic load with an allowed 10% blockage. If we wanted to provide a better grade of service, we could use 4 modems. Since we know traffic patterns are not entirely random in an e-mail situation, we'll keep an eye on the grade of service.

If the modem pool was served through a PBX, a call accounting package or PBX traffic reports on the hunt group could provide accurate traffic engineering statistics. By running a modem pool through a PBX or Hybrid system you can take advantage of traffic queuing theories to further reduce costs by sharing trunks.

A fair assumption would be that the company that has the modem pool also has a PBX or Hybrid system. This system will have CO-Both Way trunks. This system may also have DID trunks and T1 circuits. The task at hand would be to determine if the PBX or Hybrid can support analog ports (most can), and if there are available analog ports on the switch.

As voice calls are generally random in nature, with a few busy periods, and as data calls with short holding periods can also reflect voice calling patterns - we can use fewer phone lines to support a given number of users. As the user base increases (either in terms of people and phones, or servers and modems), the number of lines required to support call volumes does not grow as rapidly.

An example of this would be a small office consisting of three people. You would need at least two lines, and you would probably need three lines to keep the number of busies at a minimum. At the other end, in a recent traffic study I saw an office of some 350 - 400 people and several asynch servers. The traffic load was carried by 4 T1s, 25 CO Both-Way trunks, and 23 DID trunks. The office was served with a P.01 grade of service with 144 trunks (and could use even fewer trunks).

The savings for putting an asynch. server through a PBX include avoidance of dedicated business lines for the server, and possibly a lower cost per minute for calls as they can be run over dedicated facilities.

If we run the server through the PBX, we may need to increase trunks to the PBX. We would need to perform a traffic study on the PBX to determine if it could handle the additional load, or if more trunks need to be installed.

Assuming business line costs $38.75 per month, and the e-mail server from the previous example required 4 lines, the potential cost savings are:

$38.75 x 4 = $155 per month, or $1,860 annualized.

This is in addition to the savings we already enjoyed by changing the way callers get to the server. We were able to reach an annualized savings of $1,449.88 plus $1,860.00 or $3,309.88 by running our server through the PBX and changing from switched access to an 800 number.

The following page visualizes the project noted above.

## Phase I
## Comm Server using 4 business lines
## Callers dial Long Distance to get to server

4 Dedicated business lines to Telco

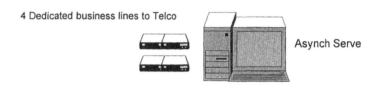

Asynch Serve

## Phase II
## Comm Server using 4 analog ports from PBX
## Callers dial 800 number to get to server
PBX

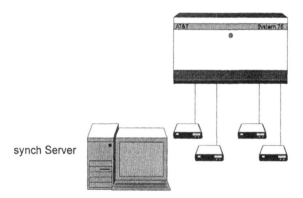

synch Server

**Savings: offset 4 dedicated trunks ($38.75 x 4 x 12 = $1,860 per year.**

receive lower cost per minute by dialing over

dedicated facilities using an inbound 800 number.

While this is a minor application as far as cost savings are concerned, the same theory can be applied to a router serving multiple dial-up clients. These devices include terminal servers, routers serving asynch clients, and routers serving ISDN BRI clients.

We looked at a simple example for a 4 port system. The savings can increase substantially as we look at servers in the 8 to 16 port size, and even larger 32 to 48 port systems.

Following is an example from a Cisco 4700 router with 2 ISDN PRIs serving multiple Cisco 1004 routers using ISDN BRI. PPP authentication chap is used to secure access. In pricing the ISDN PRIs, we also priced costs to call the central router.

We saw that it was less expensive to front end the Cisco 4700 with a pair of 800 numbers. One 800 number is for Cisco 1004s providing backup for Cisco 2500s on a frame relay network. The second 800 number is for a pilot telecommuting project.

Summary Report for a # by State - Telecommuter's 800#

| State | Date Range | Attempts | Comp | Incomp | Other |
|-------|-----------|----------|------|--------|-------|
| Florida | 07/01/96 - 07/31/96 | 3 | 0 | 3 | 0 |
| New York | 05/01/96 - 05/31/96 | 8 | 8 | 0 | 0 |
| | 06/01/96 - 06/30/96 | 166 | 166 | 0 | 0 |
| | 07/01/96 - 07/31/96 | 31 | 31 | 0 | 0 |
| | 08/01/96 - 08/28/96 | 80 | 80 | 0 | 0 |
| Oklahoma | 06/01/96 - 06/30/96 | 6 | 0 | 6 | 0 |
| SUBTOTAL | | 294 | 285 | 9 | 0 |

Thursday September 5, 1996 — Page 1 of 1

Grade Of Service - Telecommuter's ISDN Router

Thursday September 5, 1996 — Page 1 of 1

| Date Range | Grade of Service | Total Minutes |
|-----------|------------------|---------------|
| 05/01/96 - 05/31/96 | 0.0 | 122.77 |
| 06/01/96 - 06/30/96 | 3.5 | 1584.50 |
| 07/01/96 - 07/31/96 | 8.8 | 987.70 |
| 08/01/96 - 08/28/96 | 0.0 | 1377.40 |

At first glance, it looks like we may have a problem providing service in Florida and Oklahoma.

Since we haven't placed ISDN BRI routers in Florida or Oklahoma, it is expected that there would be no traffic from those states. If there WAS traffic from these states, we would have a security problem!

We also see a grade of service that is less than desirable, but we are providing a perfect GOS in New York. It looks like the incomplete calls from Florida and Oklahoma are throwing of our reported GOS.

We could also use an ANI report on CD ROM to verify where the calls are coming from. If we were charging back to pay for the 800 traffic, and marking up to pay for the ISDN PRIs and the Cisco 4700 router at the central site, the ANI report would give us the calling numbers.

We could use the ANI report on CD ROM to tie a calling number with a remote router and client. This would tell us who to bill for non-roaming clients. For roaming clients we would need to look at a different alternative for billing. In the case of a company providing access to employees, we could simply eat the cost or build a charge-back system.

We would need to bring traffic levels up substantially to break even on this project. The current traffic level will not pay for the ISDN PRI. An alternative would be to start out with a few ISDN BRIs at the central site, then upgrade to PRI as traffic intensifies.

In the case of the underutilized router with 2 PRIs, the network auditor or network optimization specialist would have a few alternatives. The alternatives include:

> Disconnect one ISDN PRI
> (at a savings of $1,200 - $1,500 per month, or $14,400 - $18,000 annualized)
>
> Disconnect both ISDN PRIs
> (at a savings of $2,400 - $3,000 per month, or $28,800 - $36,000 annualized, and install 4 to 8 ISDN BRIs (at the cost of a BRI card and 4 to 8 BRIs from the Telco, or appr. $2,400 annualized for 4 BRIs plus installation )
>
> Increase utilization through marketing enhanced network connectivity

The ISDN PRI costs are simply generalizations. You would need to review circuit billing to identify costs associated with your ISDN PRI and determine actual cost savings opportunities.

## Internet Service Provider - ISDN

If this was for an ISP project, we could possibly use 900 numbers on the ISDN PRI's. The ISP could recover costs without having to build a collections and billing staff. The client could enjoy the mobility of supporting roaming users without having to keep a book of local access numbers.

The remote ISDN routers or bridges could dial in to the serving router over ISDN BRI at speeds of 64K to 128K. With compression algorithms, we could provide speeds of up to 512K.

By using a multi-protocol router, we could create a system that would let clients call just one number no matter where they are. They would also be able to call that one number to reach a Novell server running IPX/SPX protocols. In the same call, they could reach a Unix server running TCP/IP protocols. The client could also reach an IBM mainframe that is front-ended by TN3270.

In effect, we would be providing a network service that looked and behaved the same for clients, regardless of where they are physically located, or what protocols they are running on the client side. The clients could even use the same 800 or 900 number to reach their company Intranet or to reach the Internet, by connecting the client's business router to the ISP router through a firewall.

As an ISP, this environment could be provided by installing an access router with multiple ISDN PRIs for dial-in clients. The access router would connect to an Ethernet segment on the ISPs premises. The ISP could also have a WAN router. This router would connect to the Ethernet segment the access router is on. The WAN router could have 2 types of WAN connections. One connection could be to the Internet, providing WWW, Telnet and FTP access. A second connection type could be to client businesses.

The ISP would provide two essential types of connectivity for clients. One type of connectivity would be Internet access for WWW, Telnet and FTP. The other type of access would be secure Intranet access - giving clients an ISDN dial-in solution for TCP/IP, IPX/SPX and AppleTalk protocols for local businesses.

The environment would look something like the diagram on the following page.

---

# Multi-Protocol Dial-On-Demand Routing
## Using ISDN BRI On Client Side, ISDN PRI on Serving Side
### Frame Relay to Client's Computer Center and Internet

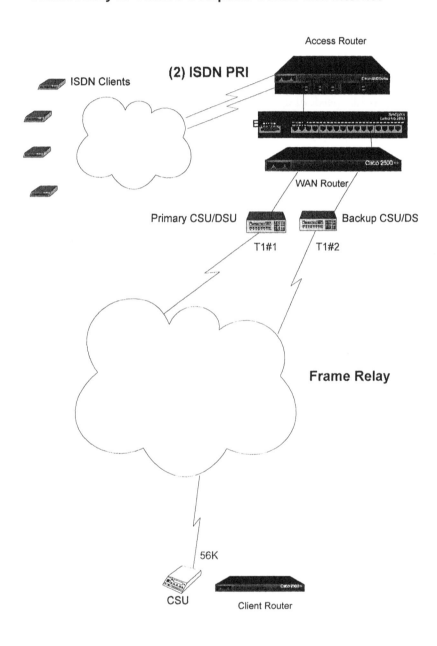

We can apply a few principles to this application: traffic queuing theory, IPX spoofing, and dial-on-demand bridging/routing.

Traffic queuing tells us that we can use fewer lines than clients as calls are random in nature (i.e. 23 PRI channels can easily serve 48 or more BRI dial-on-demand clients).

IPX spoofing tells us that we can fool the NetWare server and client into believing there is a connection up, even when the connection is down.

Dial-on-demand allows the client router or bridge to call the serving router or bridge when traffic is being generated. You can tell the router what traffic is valid, and whether it should simply receive incoming calls - or have the serving router call the client routers.

Dial-on-demand emulates voice traffic more than it emulates traditional modem data traffic. A modem tends to stay connected, whether you are transmitting data or not. A dial-on-demand router drops the call when there is no traffic. As this can be set (120 to 360 seconds of inactivity is a good starting point), we can force calls to disconnect. The call is automatically reconnected when either the client or the server sends data again.

By applying all of these principles to a dial-in environment, we can build a system that connects in a seamless manner. The client will not know when the call is up, or when it is dropped. We can set the timeout to a given number of seconds, so the call will be dropped while the client is not transmitting or receiving data. If we are using ISDN, the client will not notice the amount of time required to reconnect the call. ISDN call setup and security authentication is extremely fast.

Best of all, the client can get high speed application access. This includes 64K clear channel if running through switches that support SS#7, 128K using bonding, 512K using header and payload compression. The client can also simplify laptop administration by having one number to dial regardless of location, application or protocol.

## Switched Access

The section on modem pools is a great introduction to switched access. Switched access is a dial-up connection, that stays up as long as the end nodes are communicating. After the session is complete, the call is dropped.

This is very different from dial-on-demand routing. Bridges or routers that use switched access as a backup mechanism may not automatically drop the call when no data is passing.

While this may seem trivial, a network device left on dial back-up using switched services over a weekend can break the budget quickly! Assuming a cost of 12.5CPM, this amounts to 7.50 per hour. A device left up on call for 24 hours can eat up $180.

A device that supports dial-on-demand routing would only be up when data was passing over the circuit. That means that the call would automatically drop, even if the network technicians went home for the weekend without letting the weekend staff know about devices left on dial back-up.

Switched access can be in the form of a switched 56K circuit. The cost elements in a switched 56K circuit include circuit costs and connectivity or per minute connect costs.

Switched access can also be performed over ISDN Basic Rate Interface as noted in the previous section. The cost elements of an ISDN BRI connection include telco charges for the ISDN BRI lines, and a per minute connect cost.

Switched access can be performed on the client side using ISDN BRI, and on the serving side by using ISDN Primary Rate Interface. The cost elements in this type of connection include the telco charges for an ISDN BRI connection on the client side, costs for ISDN PRI on the serving side, and per minute connect costs.

## Leased Lines

Leased Lines can operate at a wide range of speeds. Devices can communicate over 2400 and 4800BPS circuits. A high percentage of IBM Control Units operate over 9600 BPS circuits.

Leased lines are used to form a permanent connection between two communicating end-points in a wide area network. Therefore, every leased line must terminate in 2 locations. Furthermore, the leased line must connect to 2 pieces of network hardware.

Leased lines are widely used in legacy networks, connecting IBM SNA/SDLC devices. Leased lines are also used in banking networks, and for connecting routers and bridges.

As a leased line can only be used to connect 2 end-points, more clients are moving towards frame relay. Leased lines are still prevalent in legacy networks, though. Leased lines are also cost effective when a network consists of less than 4 locations.

As the face of computing has changed, the use of 56KB circuits has also changed. 56KB circuits are often used for high volume character-based applications, such as IBM CICS regions. While a 56K circuit can support a relatively large number of transaction processing terminals, a 56KB circuit is relatively slow for LAN to LAN communications.

Higher speed leased lines include fractional T1 (128KB, 256KB, 512KB, 768KB, and 1024KB), and full T1 (1.544MBPS).


## T1
T1 service provides the building blocks for high volume, high bandwidth voice, data and video networking. While ISDN PRI, B-ISDN and ATM receive a lot of press, T1 continues to get the job done in a cost effective, highly reliable manner.

T1 is a digital transmission standard. It is used to supply up to 24 channels of voice communications to a Hybrid or PBX over a single circuit. T1 is used to build a wide area network at speeds of up to 1.536MBPS in the US. WAN devices can be routers, gateways, bridges or Front End Processors. T1s can also be attached to channel banks, providing up to 24 modem ports over one circuit.

A T1 circuit is a digital circuit operating at 1.544 Mbps. As signaling is required, the effective rate of a T1 is 1.536 Mbps. A T1 circuit can operate as a leased line, connecting 2 end points in a data communications network. A T1 circuit can also serve multiple communications applications through the use of multiplexors.

T1 provides several benefits. A T1 is often less expensive than several lower speed circuits or voice lines. While there are several rules of thumb, the only valid way of determining cost-effectiveness of a T1 is to compare prices among alternative services. A T1 circuit is normally provisioned over copper, but T1 can also be provisioned via satellite, microwave, or fiber optics.

The T1 circuit connects to a multiplexor. The components of a mux include a channel service unit, a multiplexor, and port(s) to connect to network equipment such as PBXs, Hybrids, Routers, Bridges and Front End Processors.

T1 line encoding can be either AMI or B8ZS. When selecting one signaling protocol over another, it is important to make sure that the CSU is optioned to the same format that the T1 is provisioned with. It is also important to make sure that the communications equipment is optioned the same.

AMI signaling offers speeds in multiples of 56K. B8ZS signaling offers clear channel, or speeds in multiples of 64K. This is not very important at slower speeds; 56K versus 64K, 112K versus 128K.This becomes important at higher speeds, such as 224K versus 256K, or 448K versus 512K. If the network equipment and T1 multiplexors in place can support B8ZS, it is better to use B8ZS for data communications.

A wide variety of T1 multiplexors are available. Some support analog ports, for connections to modems, or analog trunks on older PBXs. Some support several digital ports, breaking a T1 up into 24 DS0, or 56K circuits. Some multiplexors support various circuit speeds (9.6KB, 14,4, 19.2, 56K, 112K, ...).

A large portion of multiplexors in use are drop and insert muxes. They may support only two ports. One port could be used for data communications, and the second port could connect to a T1 card on a PBX or Hybrid system.

T1 circuits have played a role in larger offices, and are very common in large offices that have PBXs. T1s have also come into play in Hybrid systems, with some degree of luck (don't ask me about a specific vendor that offers T1s on Hybrids, but does not have a large pool of technicians trained in T1s, and the technicians arrive to test T1 circuits without T1 test sets).

As LAN-to-LAN communications even in small offices are starting to require Fractional T1 service, it can be cost effective to look at a T1 mux that provides one V.35 port for a frame relay network, and 6 analog ports for a line pool on a KSU.

Small offices do not normally require the voice communications features and functions of a PBX. A large Hybrid may offer more than is required, at a higher price. Small offices can be well served by KSUs. If a small office requires Fractional T1 for data, it could be cost effective to add channels for voice, especially as the T1 access charges are already covered under the data circuit.

This would allow smaller sales offices to participate in a virtual private network offering, and to enjoy the benefits of reduced long distance costs. While only a portion of the T1 would be used, the data portion could be muxed out at a DACs at the vendors Point Of Presence (POP).

The voice could also be muxed out, providing on-net calling and dedicated access for low cost Long Distance dialing. Integrated Fractional T1 service combining voice and data could increase a carriers presence in smaller clients voice and data networks.

**Frame Relay**
Frame Relay is a packet switching offering for wide area network communications. Frame Relay is offered by many carriers, including AT&T, MCI, Sprint, WorldCom (WilTel), several (if not all) RBOCS, and several secondary carriers.

Frame Relay is used to connect routers and bridges that are in different geographic locations. Frame Relay can also be used with newer frame capable IBM devices. Through the use of FRADs, frame relay can be used to connect devices that normally communicate over point to point leased lines. This allows older legacy network equipment to be converted to a frame relay networking environment.

Frame Relay devices connect to the carrier through an access loop, or circuit. The access loop can be a 56K DDS circuit, or it can be a T1 circuit. While Fractional T1 speeds are available, the circuit still connects via a T1 access loop and a CSU.

The access circuit is normally provided by the local Telco. This would be Nynex, Bell Atlantic, Bell South, Southwestern Bell, US West, or PacBell. This could also be GTE. In areas where there is competition in the local loop, the loop may be provided by an alternative carrier. This could be MFS (now part of WorldCom), ICG, or others.

As in the case of a leased line operating over a 56K DDS circuit or a T1, the circuit comes in to the Dmarc. It is then extended out to the client's computer room, either by the Telco (at a cost), or by the client. In the case of a T1, the circuit is then terminated in a smart jack. Some smart jacks are programmable, and need to be set for proper signaling.

On the customer premises, the circuit is then connected to a CSU/DSU. This can be a 56K CSU, or a T1 multiplexor with a built in CSU/DSU. This piece of equipment can be provided by the client, or it can be leased or rented from the frame relay vendor.

If the CSU is provided by the client, then the client is responsible for optioning the CSU. If it is provided by the frame relay vendor, then the vendor is responsible for the frame relay portion of the network, the local loop (provisioned through the local Telco or an alternate access provider), and the CSU.

Billing starts at one of two dates, depending on who provides the CSU and accepts the local loop. Billing can start when the loop is in if the client provides the loop. Billing can also start when the CSU/DSU is installed, optioned and tested if the vendor provides the CSU.

**Frame Relay Billing**

The billing portions of a frame relay circuit include:

**One Time Charges**

Local loop installation

Extend Dmarc to computer room

Frame vendor's Port installation charge

**Monthly Re-Occuring Charges**

Local loop, or access

Port charge on vendors network

CIR

PVC/DLCI

**Optional Equipment & Management Costs**

CSU installation

CSU monthly rental or lease

Router installation/configuration

Router monthly rental or lease

Network Management

Costs are variable. The local loop is mileage dependent. In pricing frame relay, or in auditing frame relay - it is a good idea to review applicable tariffs.

The port charge is speed dependent. There will be a lower price for a 56K port than there would be for a higher speed port. In auditing the frame relay bill, it is important to verify port speeds and compare costs.

If a business has several frame relay circuits operating at the same speeds, the access rates would vary (remember, access is mileage dependent), but port charges should be the same for ports of the same speed.

Frame Relay devices communicate with other frame relay devices through DLCI (or Data Link Connection Identifier) mapping. This is also known as PVCs, or Private Virtual Circuits. Carriers bill on PVCs, therefore it is important to verify that PVCs or DLCIs are in place, are communicating with each other, and are passing data to each other.

A business may have built a fully meshed frame relay network. This would require each office to have a PVC mapped to every other office. This may not be the most cost-effective way of providing communications across the network, as some offices may have little or no need to communicate with each other.

Carriers often over-sell network meshing. The term "fully meshed network" sounds impressive. The bill for a fully meshed network is certainly impressive. Meshing requirements need to be determined based on where the clients reside and where the servers reside.

Communications requirements are based on application requirements. This becomes apparent as the auditor verifies where the communicating end nodes are located. The network simply provides a pipe. It is important to verify where the applications servers reside, and where the clients reside.

If a business has a central operations center, and a few inter-connected LANs, you may find that communications is largely local to the LAN, with intermittent access to the central operation center. Even if offices communicate to each other via e-mail, the e-mail servers may reside at each office and the e-mail routers may reside at the central operations center.

In auditing the network, you would review what PVCs are on the bill, and where the PVC end points are. You would then use a router's command line interface to make sure that the PVCs in a fully meshed network are sending information to network devices. An alternative would be to build a network auditing tool kit, further described in Appendix A.

This tool kit would give the auditor the ability to place a PC on a client's network and trend wide area network utilization over a period of time. While the cost for the tool kit may be somewhat high, the payback period could by relatively quick. This would be analogous to placing a polling device on a PBX and providing a service bureau for traffic engineering.

An alternative is to use the frame relay providers reports. This will show utilization, which you can use to determine the appropriate port speed. This will also show dropped bits, which can help you determine the appropriate CIR. These reports will also show traffic by PVC, which can help you determine if you can disconnect a PVC in a meshed network.

## ISDN BRI

ISDN Basic Rate Interface is used as a dial-up access method for data communications. ISDN BRI is often compatible with Switched 56K service. BRI can operate at 64KB, or 56KB.

For a call to go through at 64K, the call must be from an ISDN BRI device, to an ISDN device, and communicating across central office switches that support ISDN using SS#7 signaling. A call will be made at 56K if the call goes through switches that do not support ISDN BRI.

Several network devices support ISDN. Some of these devices are bridges. Some devices are routers. Some act like modems. It is critical when considering ISDN BRI as a transport, to also consider what devices communicate at the client end, what devices communicate at the server end, and how the devices communicate.

It doesn't make sense to purchase an ISDN modem if the application requires a bridge or a router. It also doesn't make sense to purchase an ISDN BRI router if you are communicating with an online service that expects a modem or terminal adapter (TA).

The ISDN BRI service elements that the data communications auditor needs to be aware of include: location, ISDN phone #1, ISDN phone #2, SPID #1, SPID #2, Circuit ID, CO Switch, and profile. The auditor will also need to know the monthly rate for ISDN service, installation costs, and the cost per minute.

The help desk and network staff will need to know who to call if there is a problem with the ISDN circuit, and where to circuit is located.

When working with ISDN dial-up services, the auditor will need to compare the cost per minute for dial-up service. ISDN can cost between 12 cents per minute and up to 30 cents per minute for a data call. This is dependent on where you are calling from, and what the tariffs are.

An alternative is to set up an ISDN pool where you will be dialing in, and assign an 800 number to it. This way you will be charged for the 800 service, which is often lower than a normal long distance call. This becomes especially attractive if the network equipment supports data over voice, allowing you to enjoy lower rate voice calls with data equipment.

## ISDN PRI

ISDN PRI is the Primary Rate Interface. This operates over a T1 and gives 23 channels of bearer service, and 1 channel of data for signaling. The PRI can be split out to support BRI. This can be done through a mux. Some routers and bridges also support ISDN PRI in the serving node, with clients calling in using ISDN BRI at the client side. The router multiplexes the calls out over the PRI.

ISDN provides the basic building blocks for computer telephony, video conferencing, and dial-on-demand data communications. ISDN BRI is appropriate for smaller scale communications requirements. ISDN PRI is appropriate for larger volume traffic.

ISDN PRI is used for voice communications on an ISDN PBX or Hybrid. ISDN PRI is used for dial-on-demand data communications, offering up to 23 simultaneous connections per PRI. ISDN PRI is also used for Video Conferencing, using H0 standards (6x64K, or 384K connections), and nx64 standards.

The signaling channel can be shared across multiple ISDN PRIs. This gives you the ability to support 24 channels on a PRI, with the signaling going over a separate PRI.

In a data communications audit you will primarily find ISDN PRI used as a dial-in platform for disaster recovery and for telecommuters. You may also find ISDN PRI used for video conferencing. The key points to be aware of with ISDN PRI include:

> Access cost for local loop
> D-Channel signaling cost
> Use (voice, data, video)
> Number of clients
> Connecting server     (router, gateway, bridge)
> Clients     (routers, bridges, terminal adapters)
> Cost per minute for dial access

Some Telcos charge more per minute for data ISDN calls than for voice calls over ISDN. In this case, the auditor should verify cost per minute for call type and verify if the clients and server can follow the Data Over Voice standard. This will drop the speed of the call from 64K down to 56K, but can improve the bottom line.

In one instance we noticed a 25 cent per minute per channel charge for data calls over an ISDN PRI. Voice traffic over an 800 number cost 8.3 cents per minute. Our target was 33-40K minutes per month, and our current was about 8K minutes per month.

Data Over Voice Alternative Pricing
8,000 minutes at 25CPM        = $2,000.00
8,000 minutes at 8.3CPM       =   $664.00
This generates a monthly savings of $1,336, or an annualized $16,032.

Using our target of 33,000 minutes per month, this comes out to:
33,000 minutes at 25CPM       = $8,250.00
33,000 minutes at 8.3CPM      = $2,739.00

At our target, we would see a savings of $5,511.00 per month, or an annualized $66,132. Not bad for dropping our speed down from 64K to 56K!

## Cost Analyses - Circuit Switched Communications Versus Leased Lines

The network auditor or network optimization specialist may read articles on circuit switched data and the promise of ISDN Basic Rate Interface and Primary Rate Interface. While ISDN allows quick call set up and requires less manual labor on the part of Telcos, ISDN also bears a cost per minute for usage.

For network devices that do not support dial-on-demand routing and bridging, protocol spoofing, and automatic disconnection when no traffic is being transmitted or received - leased lines (and frame relay) can be very attractive.

The points to review in a cost analyses include:

> what devices need to communicate with each other
> how much time per month will the devices be communicating
> do the devices support dial-on-demand, or is the call manual
> will the devices drop the call when nothing is being transmitted,
>       or is the call dropped manually
> what is the cost per minute for switched access (analog, ISDN, ...)
> how many minutes per month will the devices be communicating
> what circuit speed is required (either dial-up or leased)
> what is the installation charge for a leased circuit
> what is the monthly re-occuring charge for a leased circuit

As an example, let's assume that we have a LAN in a small office, and a corporate LAN at headquarters. Let's assume that we have an e-mail server at the small sales office, and an e-mail router at the headquarters. Let's also assume that we have some LAN to LAN communications requirements, and possibly a need for access to mainframe applications.

If the sales office had a small staff, and was using a router that supported dial-on-demand routing and dropped the call on inactivity, we could estimate costs. An estimate might be:

> Average of 20 minutes per hour application WAN access
> Average of 160 minutes per day
> Average of 3,200 minutes per month
> 56K to 64K speeds acceptable, therefore 1B channel is okay
>
> ISDN BRI costs office $72.83 per month
> ISDN BRI costs office 12 cents per minute
>
> $(3,200 * .12) + 72.83 = \$456.83$ per month, switched

This compares favorably with 56K frame relay from a majority of vendors, therefore we may chose an ISDN dial-on-demand solution. We may want to hedge the bet as follows:

> 56K frame relay from vendor A costs $520 per month
> Headquarters does not require additional bandwidth

If the sales office increases monthly connect times by 527 minutes per month, it's a wash between ISDN and 56K frame relay, provided there is capacity on the headquarters location frame circuit.

We may want to trend connect times for three months. This will tell us if we should migrate the small office off ISDN and on to frame relay.

Frame relay installation costs and CSU costs will need to be taken into effect to determine the payback period. Cost savings can be calculated after the payback period is met.

If the office can not support any network down time, we may want to keep ISDN in place to provide backup. This would give us disaster recovery for the small office.

# Part III - The Current Network

## - The Interview Phase

Your interviews will help you learn a lot about the network -- both as it exists today, and as it is perceived today. You will want to interview people from each networking discipline, as they will have a different view of network components and network connectivity. Each group may or may not use different documentation tools, and these tools may or may not be integrated.

Carefully think about what you want to accomplish before scheduling interviews. Network staffs are typically understaffed, and over-stressed. Help Desk and Network Control Center people only get called if there is a network outage, and they may be getting called quite often!

Schedule your meetings around times that fit best for the groups you will work with. Do not try to schedule a meeting with front line people on a Monday morning. They will typically be very busy. You will have better luck meeting in the afternoon in the middle of the week, or meeting during the second shift.

While interviewing each group, ask them what information systems are used to identify network components. Ask specifically how to log on to the information system, and how to make inquiries. You will probably need a userid and password. It is best to ask for these as your project gets accepted.

### Network Design Team
The network design team is a great place to start in determining how the network was originally designed, and what the goals and objectives were. You will be able to determine what the communicating devices were, and what protocols and speeds were originally required.

The network design team is not always involved in day-to-day activities of the network. As networks evolve, they sometimes do so without the guidance of a designer. This makes it important to meet with Network Control and Help Desk personnel.

## The Help Desk

The Help Desk is the front line. They receive calls whenever there is an outage. The Help Desk also receives calls when response times are poor. Normally calls are logged in a Help Desk application.

By gaining access to the Help Desk application you will be able to search for problem records on the network you are auditing. You should be alerted if there are excessive problem calls from a specific office – it may mean that there is a reliability problem with networking components at that office.

You should also be alerted if there are very few, or no problem calls at all from a specific office. They either are having uncommonly good service, or are a prime candidate for a disconnect.

While speaking with Help Desk personnel, you need to find out what the network problem tracking inquiry is. Ask what the Help Desk technician asks the remote office when a problem is called in. Some of the basic information would be: where are you located, what is the name of your network equipment (network name, router name, ...), where are you trying to go, has it ever worked before.

By learning how to use the Help Desk's information system, you will get a good idea of what equipment the Help Desk knows about.


## The Network Control Center

The Network Control Center typically handles calls that the Help Desk can't clear. While interviewing the Network Control Center, you will get a better picture of what the network is like.

The Network Control Center will have a management information system that is shared with the Help Desk. This system should let you know where offices are located, what wide area equipment is in use, and what circuits are used. This information will also be incomplete, inadequate, and incorrect.

Information Management systems are often poorly maintained. The day-to-day routine of handling problem calls and resolving network problems leaves little time to maintain an information system. While information is critical to the success of a Help Desk and a Network Control Center, you will find that both groups are understaffed and do not have the time to maintain information systems.

While working with the Network Control Center, ask what type of calls they handle, and how they identify a remote office. Ask what information is critical to problem resolution, and what information a caller normally has.

This can lead to a great second phase of your consulting / contracting assignment if you haven't already priced it out! The phase would be documenting the network in the Help Desk's / Network Control Center's information management system.

It would also include sending procedures out to each remote office on how to report problems, and what information is required to identify the remote office. This should be included in procedures you would provide for remote offices.

### Network Analysts and Network Billing Personnel
In larger networks you may find people that specialize in provisioning circuits and in network analyses. If your company has these people, it is critical to meet with them and understand what they do.

A network analyst should have a good idea of network utilization. This will help you determine where speeds need to be modified. The scope of this book does not include network analyses and design, nor does this book deal with the required tools.

Your research will give you a good idea of what is required, but design tools such as NetSuite can be used if you plan to go further than auditing the network. Alternatively, reports from your network auditing tool kit can be helpful, especially if you use utilization trending applications.

# - Bills, Invoices, Docs

### Obtaining Bills and Invoices
One of the most difficult tasks in documenting and price engineering a network is to obtain accurate bills. Bills are available in several forms and formats.

If services are provided by a Regional Bell Operating Company, data communications circuits may be billed under your company's telephone bill. This is often difficult to break out. Some invoices don't even break out circuits and connecting points, they are simply billed in the same manner that a CO-trunk is billed.

Billing detail is available when you contact your marketing representative and request a copy of your Customer Service Records. Customer Service Records are a critical component in auditing a data communications network.

While the format of the CSR may look a little daunting, it is actually fairly straight-forward. "The Complete Guide To Local & Long Distance Telephone Company Billing" is a great reference.

If your network is provided by an IXC such as AT&T, MCI, Sprint or Wiltel, the network bills will look different. The bills will reflect the circuits that you have, the end points, PVCs, CIR, and port speed. While the look is different across carriers, the information will be the same. Therefore it is important to know the items that make up a network bill, so you know what questions to ask your carrier.

Many companies are moving towards Electronic Data Interchange. EDI enables companies to issue electronic purchase orders, shipping documents, and invoices. EDI shortens the payment cycle as bills are paid electronically when they are received.

While EDI can improve the process, it can also make it very difficult to properly track what items are being paid for, and if the correct price is being paid. EDI also makes it difficult to track disconnect orders and change orders. Paper copies should be kept on file.

If the company you are auditing uses EDI billing, you will need access to the system. You will need either a userid and password, or printed reports.

### Reviewing Bills and Invoices

Network bills should be reviewed every month. It is a good practice to keep a copy of all network bills, and to track them monthly.

As you perform cost-savings activities, it is imperative to make sure that any cost savings initiative actually follows through to the bill. If you issue a circuit disconnect, review the bills for the next two months to make sure the circuit is no longer being billed to you. If you issue a request to reduce circuit speeds, make sure that the request is completed and that the monthly bill is reduced.

One good method of tracking costs is to keep costs in a spreadsheet. If your carrier does not offer billing detail via diskette, this can be a manually intensive effort. The effort is worth while, as you will be able to track monthly expenses, and begin to trend network costs.

This will put you in a good position to start a budget for next years communications requirements. This will also help you determine if you are in danger of going over budget.

Alternatively, network costs could be manually (or copied via file transfer) tracked in a telemanagement system. This would enable the network group to document network costs by location, and can provide a basis for building a charge-back system. The charge back system could include both voice and data network costs.

### Customer Service Record (CSR) For T1

In a previous section on verifying router ports, we came across a T1 circuit that was connected to a router at customer premises A. The circuit was not connected at customer premises B.

The T1 was provided by the local Telco. Following is a copy of the CSR. It is interesting to note that both customer premises A and B are served by the same CO. The T1 has a local loop charge on both legs, but does not have a mileage charge.

In looking for the T1 costs, we did not have a bill that specified the circuit. We did have a very large monthly phone bill with a large fixed portion. The following T1 was in the phone bill, billed to the main listed number.

```
******. DHDA.994168..SN. *********************************
      CLS    .DHDA.994168..SN./SSP
      CKL    1-38 MAIN STREET, DANB, COMP RM
      CKL    2-23 RIDGEVIEW DRIVE, DANB, COMP RM
      SCS    HCAXL                              ITEM RATE  TOT CHRG
             1 HCAXL                                  NR        NR
             1 1LDPX/REF 1/DES CKL 1-DANB      ($nnn.nn)
                /LSO 203 796/BTO
                203  471-1378/CUS 569
             1 1LDPX/REF 2/DES DANB-CKL 2      ($nnn.nn)
                /LSO 203 748/BTO
                203  471-1378/CUS 569
             1 RJ48S/CKL  1/LSO 203 796/BTO              .00
                203  471-1378/CUS 569
             1 RJ48S/CKL  2/LSO 203 748/BTO              .00
                203  471-1378/CUS 569
```

CKL-1 and CKL-2 are locations 1 and 2 of the circuit. This tells us the two end points that are connected together. The circuit ID also appears on the CSR. This should correspond with a circuit ID written on a block at the Dmarc.

If we were to visit the premises, we should be able to track circuit DHDA 994168 from the Dmarc to a CSU. The CSU should also be labeled with the circuit ID if it was provided by the Telco. Finally, we should be able to follow the V.35 cable connecting the CSU to a piece of networking equipment.

In this case we found it connected at premises A, and disconnected at premises B. Our alternatives are to connect it at premises B, or disconnect the entire circuit.

## 56K Leased Line - Billed by RBOC

Following is a Customer Service Record (CSR) for a 56K leased line billed by the local Telco. CSUs are provided by the customer at both legs of the circuit. This leased line priced out as less expensive than a frame relay solution as both ends are served by the same Telco.

This circuit was also "hidden" in a local phone bill as part o the monthly re-occurring fixed charges.

If the client had several offices within the same regional area, we may look at a frame relay solution provided by the RBOC. As the business is geographically disbursed, the company is using a combination of frame relay and leased lines.

```
***** .DWDA.923507..SN.*******************************************
CLS    .DWDA.923507..SN./SSP          HEADQUARTERS
CKL    1-38 MAIN STREET,DANBURY
CKL    2-240 ELMWOOD DRIVE,NEWINGTON,
SCS    PLDRS                          ITEM RATE   TOTAL CHRG
       1 PLDRS                        NR          NR
       1  1L7HT /REF 1/DES CKL 1-DANB   (nnn.nn)    (nnn.nn)
              /LOC BASEMENT COMM ROOM
              /LSO 203 796
       43 1LNPX/REF 2/ZIWC DANB-HFD NGTN  (MILEAGE RATE)
       1  1L7HT /REF 3/DES HFD NGTN-CKL 2  (nnn.nn)   (nnn.nn)
              /LOC GROUND FLOOR
              /LSO 860 666
```

Billed items:

    Local loop: customer premises A

    Mileage charge (1LNPX)

    Local loop: customer premises B

Audit Items:

    Verify circuit up on both legs

    Verify mileage from V&H coordinates

    Verify tariffed rate

We can check connectivity by reviewing information system records. Do we have networking equipment at location A and location B (ckl-1 and ckl-2). What is the network address at both locations? Is the circuit up and passing information?

In this case the records only show one piece of networking equipment at premises B. We dialed into the hardware at premises B and checked the serial port. We saw that it was up/up and passing information.

### ISDN BRI Customer Service Record
Following is a CSR for an ISDN Basic Rate Interface line. The ISDN BRI is being used for router network recovery, therefore it is part of the network audit.

```
NP       (SPNP) CUSTOMER NAME
LA       100 CENTURY BLVD,
         SPRNGDL  (45246)
SIC      S6141
         BILL
BN       CUSTOMER NAME
BA       38 MAIN STREET
PO       DANBURY, CT  06810
ZCC      HA
         RMKS
RMKR (A) SAG=YC
PCS

         S&E
1        DXE       /TN 9840 / PIC MCI, I      NR  965   K7777735 012296
         /LPIC MCI
         /DSNA   IBSD007466    CB
         /DES SPID = 016719840000
         NATIONAL
1        ACB2X  /DSNA  IBSD007446 CB      NR  182  N4401349 122695

2        LTQ1X                       $ 51.86  960 N4401349     122695
1        LOY  /PIC MCI , I / LPIC MCI $ 55.00  959 K7777735     012296
         /DSNA   IBSD007446  CB
1        1BQ                         NR  965 N4401349     122695
1        NP3                         NR  202 N4401349     122695
1        9ZR                      $  4.87 309 N4401349     122695
1        RJ4   /ZNI RJ48S             NR  182 N4401349     122695
1        DXE  /PIC MCI, I / LPIC MCI  NR  965 K7777735     012296
         /DSNA   IBSD007446  CB
         /DES     SPID = 016719909000
         NATIONAL

                             $  111.73  TOTAL
         ORD  K7777735 01-22-96
         2    911 CHARGE(S)       $    .24
```

This is an EXPENSIVE BRI, requiring a call to Cincinnati Bell. The items an auditor would review are:

     Bill details        - being billed properly?
     ISDN BRI        - equipment connected on premises?
     Network equipment - test dial to ensure connectivity

This CSR tells us the circuit ID for the BRI, the SPIDs, the format (National), and the carrier.

### Reviewing Network Documentation

Network Documentation varies substantially depending on network equipment, network services and the accuracy of record keeping.

Additionally, if Fractional T1 service is being used, you should have a record of the channels assigned. This will let you plan on future growth for data, or the addition of voice services over spare capacity. This information is important regardless of your carrier.

Financial information would include the cost of the port on the vendor's network, the cost of the local loop, the cost for a given CIR, and cost for PVCs.

### Documentation worksheets

The documentation worksheets on the following pages may be freely copied by the purchaser of this book.

The worksheets can be used while performing a network audit. You will find them useful, in that they can help identify the questions you should ask, and help point out the information that you should look for.

Feel free to make any changes to the worksheets, to best meet your needs, or your client's needs.

# Site Survey Form

Company Name: _____
Street Address: _____
Suite: _____
City: _____ State: _____ Zip: _____

Primary Contact: _____
Phone Number: _____
Fax Number _____
Pager Number: _____
Responsibility: _____

Secondary Contact: _____
Phone Number: _____
Fax Number: _____
Pager Number: _____
Responsibility: _____

Building Management Contact: _____
Phone Number: _____
Fax Number: _____
Pager Number: _____

# Leased Line Documentation - Fractional T1 - T1

## Identifiers
Location _____
Contact _____
Customer Order Number _____
Vendor Order Number _____
Vendor Service Number _____
T1 ID _____
Channels Assigned _____
Circuit Speed _____
Telco Local Loop _____
T1 Format D4/AMI, ESF/B8ZS, ...

## Provisioning
CPE Provider Customer OR Vendor
Date requested _____
Installation date _____
Production date _____
Dmarc location _____
Dmarc extension Telco or customer
CSU Installation date _____

## Purpose
Connecting Equipment    Router - Bridge - Gateway -  FRAD
Equipment Port _____
Equipment Install Date _____
Equipment Production Date _____

## Financial
Local Loop charge _____
Port Charge _____

Spare Channels on T1 _____
Available port on Mux? Yes  /  No
PBX support T1? Yes  /  No
T1 used for voice and data? Yes  /  No
If yes - PBX ports assigned _____
If yes - PBX trunk group _____

# Frame Relay Documentation - Fractional T1

## Identifiers
Location                         _____

Contact                         _____

Customer Order Number      _____

Vendor Order Number         _____

Vendor Service Number        _____

T1 ID                            _____

Channels Assigned            _____

Circuit Speed                  _____

Port Speed                    _____

CIR                              _____

Circuit ID                      _____

Telco Local Loop             _____

T1 Format                     D4/AMI, ESF/B8ZS, ...

DLCI                         _____

## Provisioning
CPE Provider                  Customer OR Vendor

Date requested               _____

Installation date            _____

Production date              _____

Dmarc location              _____

Dmarc extension            Telco or customer

CSU Installation date        _____

## Purpose
Connecting Equipment     Router - Bridge - Gateway - FRAD

Equipment Port              _____

Equipment Install Date      _____

Equipment Production Date   _____

## Financial
Local Loop charge           _____

Port Charge                  _____

PVC Charge                 _____

CIR Charge                 _____

## Management
Performance reports costs   _____

# ISDN BRI Documentation

## Identifiers
Location _____
Circuit ID _____
Phone #1 _____
Phone #2 _____
SPID #1 _____
SPID #2 _____
CO Switch _____
Profile _____

## Provisioning
Telco Contact _____
Telco Phone # _____
Conf. Number _____
Installation Date _____
Inst. Location _____
Jack Type (RJ48-S,...) _____

## Purpose
Use (Voice, Data, Video) _____
Connecting Equipment _____
Equip. Inst. Date _____

## Financial - Local Service
Installation Cost _____
Monthly re-occurring charges _____
Cost-per-minute, local calling _____
Average minutes per month _____

## Financial - Long Distance
Cost-per minute, first 30 seconds _____
CPM, 6 second increments _____

## Router Documentation

Location - City, State, Zip      _____
Location - Room, Distribution Rack      _____

Manufacturer:      _____
Model:      _____

Software Version:      _____
Memory:      _____

WAN Interfaces:
     Serial Ports      _____
     TCP/IP, IPX/SPX Addresses      _____

     ISDN BRI ports      _____
     TCP/IP, IPX/SPX Addresses      _____

     ISDN PRI ports      _____
     TCP/IP, IPX/SPX Addresses      _____

     ATM Ports      _____
     TCP/IP, IPX/SPX Addresses      _____

LAN Interfaces:
     Ethernet ports      _____
     TCP/IP, IPX/SPX Addresses      _____

     Token Ring ports      _____
     TCP/IP, IPX/SPX Addresses      _____

     Fast Ethernet ports      _____
     TCP/IP, IPX/SPX Addresses      _____

     FDDI Ports      _____
     TCP/IP, IPX/SPX Addresses      _____

# Router - Administrative Detail

Router name:          _____
Location:             _____
Office hours:         _____


Contact Name:         _____
Phone Number:         _____
Fax Number:           _____
E-Mail Address:       _____
Pager Number:         _____


Maintenance modem attached? Y/N      _____
Asynch access number:                _____
Browse password:                     _____
SNMP Read community string           _____
SNMP Write community string          _____


Manufacturer:                        _____
Model:                               _____
Serial Number:                       _____
Maintenance Contract Number:         _____
Support Plan                         24x7, 8:00-5:00, Drop-Ship
Manufacturer Support Number:         _____

# - Hitting The Road

### Site Surveys - The Go or No Go Decision

As you go through network bills, you may find that you need to go to a remote office for further information. Before deciding to hit the road, review the network bills and network documentation.

You will often be able to determine which circuits are used to connect which communications end-points, simply by learning where the circuit originates (point A), where it terminates (point B), what the circuit speed is, and who the carrier is. This is especially true in smaller remote offices that communicate with one or two other locations.

A large office with several communications circuits has a greater opportunity for savings. As the network grows, older circuits may or may not have been disconnected. New circuits may have been earmarked for that new office - which never did get put on the network. You will want to visit a large office and perform a true site audit.

### Who Pays for T&L?

Before hitting the road, determine what the company's travel and living policy is. If you are acting as a consultant, you will want to bill T&L back to the client. Determine up front if the company has policies and guidelines on the type or price range of hotels you will stay at. Determine up front what the allowed expense is per day for T&L.

Before making hotel arrangements, ask if the company has a relationship with any nearby hotels. Compete the rate you are charged versus the rate the company receives.

### Verifying 56K leased lines

How do you verify a 56K circuit? A 56K circuit normally comes in on an 8 wire, 4 pair cable. It will normally come into the building on an RJ21X block. Either the telco or the client will have extended the D-marc to the client's computer room. It will be terminated in either an RJ45 or an RJ48 jack (both jacks look the same).

If you have any luck at all, the circuit will be labeled on the RJ21X. It should also be labeled at the RJ45 or RJ48 jack in the computer room.

The 56K circuit will connect to a CSU. The connection will normally be an 8 wire RJ45 cable, extending from the RJ45 jack to the back of the CSU. The CSU will have an interface on the back of it to support data communications equipment. You will normally find a V.35 interface on the back of the CSU.

Following is an example of a 56K circuit.

Punchdown Block

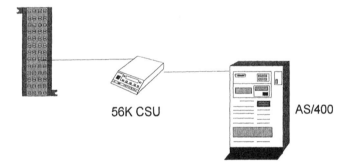

56K CSU          AS/400

Circuit is extended from punchdown block to CSU.

Circuit connects to CSU using 8 wire connector, looks like RJ45

CSU is connected to network equipment using a V.35 cable

**Verifying Fractional T and T1 lines**

There is no difference in wiring between a Fractional T1 (FracT) or full T1 circuit. It comes in to your building the same way that a 56K circuit comes in.

Rather than terminating in a 56K CSU, your T1 will terminate in a T1 CSU/DSU. You will find that the T1 circuit normally comes in on the RJ21X, which should be labeled. You should see a cable that is punched down on the RJ21X.

The cable can is connected from the punchdown block to the telco provided CSU. This is often done using a modular RJ45 connector. The circuit is then extended from the telco provided CSU to the client (or vendor) provided T1 multiplexor.

To verify your T1, simply go from the RJ21X to the telco supplied CSU. Look at the lights. You should have a green light indicating good power. You should have a light that indicates either ESF or AMI. If this is clean, then your T1 is part-way there!

The telco supplied CSU is connected to your carrier's T1 Multiplexer by way of an 8 wire, 4 pair modular cable in most cases. If the telco supplied CSU is in a different room than your carrier supplied CSU, the inside cable will have been extended to your communications room. The T1 multiplexer will have lights on it indicating power, signal and formatting.

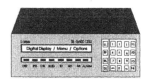

## Verifying Hardware Connectivity
### The CSU

We already discussed how the circuit extends from the RJ21X block to your CSU. The back of your CSU or T1 multiplexer will have one or more interfaces.

In the case of a 56K circuit, the interface is used for data. You will normally have a V.35 interface on your CSU. This will connect to your communications equipment by way of a V.35 cable.

In the case of a T1 circuit, you may have multiple interfaces. These can be used for data, they can be used for voice, or they can be used for a combination by using a drop-and-insert mux.

### The Router

Routers will be the most common piece of wide area networking gear in the near term. You may also run across bridges, but bridge use is waning - except as a low-end dial on demand platform for ISDN networking.

Please refer to Harry Newton's Telecom Dictionary and to Christine Heckart's "The Guide to Frame Relay Networking" for an in depth look at routers and routing protocols. (See the last page of this book.) You should also visit several Web pages, including Cisco's and Bay Networks. You may also want to look at the various router and networking newsgroups.

We are simply interested in the interfaces on a router or bridge, and what is connected to those interfaces. For a wide area networking audit, we only need to look at the serial interfaces. For a true network audit, we will need to look at all interfaces.

While reviewing router physical connectivity, look at the circuits you traced back from the RJ21X or 66 block to the CSU (for 56K) or to the T1 multiplexer (for FracT and above). At the back of the CSU or Mux you will find an interface and a cable. You will want to trace where the cable connects on the router.

If the cable is not connected to anything (I've run across several!) make note of it. The alternatives are: the router was upgraded to a higher speed circuit, the router was moved to a different carrier, or a circuit was ordered which hasn't been used yet.

As you write up the results of your inventory - include all CSUs and T1 Multiplexors that are on site and not in use. In a future chapter we will discuss how to dial into a router or telnet to the router and check the logical status of the serial ports. This will tell you if the circuit is up and passing data, or if the circuit is not in use.

While you are at the site, you should also note the number of Ethernet interfaces, the number of Token Ring interfaces, where the interfaces are connected to a hub or switch, and if a modem is connected to the router for remote diagnostics. If a modem is connected, you will want to get the phone number in case you need to dial into the router to verify port status at a later date.


**The Control Unit**
While routers are growing in popularity, there are still several legacy shops out there. A legacy shop is a mainframe processing shop that uses IBM SNA/SDLC equipment. While SNA/SDLC equipment is waning, there are still uses for 3x74 control units, 37x5 Front End Processors, and 3178 terminals. This is especially true in transaction processing environments.

You may find leased 9.6 circuits in use to support IBM 3x74 control units, or other networking equipment. You could also find 56K circuits serving SNA/SDLC equipment.

An addition to IBM Front End Processors and 3x74 control units, you may come across IBM AS400s. There are several AS400s that meet business needs as a solid minicomputer with good networking capabilities.

For a circuit inventory / auditing approach, we are mainly concerned with what is used to connect the 3x74 control unit to the wide area network. This will either be a 56k circuit connected through a 56K CSU, or a 9.6K circuit connected to a modem in leased line mode. IBM controllers can also connect to a bridged network with a Token Ring card.

The information you should have gathered during your site survey includes:
data circuits connected to RJ21X
connection from RJ21X to CSU
(if CSU is in a rack, you should note which slot)
connection to T1 multiplexer
connection from CSU or mux to network gear
port or interface circuit is connected to
if shared access, both voice and data devices T1 connected to

modem attached to equipment for remote maintenance
dial-in number to modem
equipment modem is connected to

serial number for router / bridge / controller, ...

If you noted any equipment that was not clearly labeled, you should label it. Include circuit ID and communications equipment port. If you decide to move to another carrier, it will be a lot simpler if you can state: "please take your new cable and attach it to your second serial interface (S1). Attach the new cable to the Brand X T1 multiplexor. After your new circuit comes up, remove the cable from S0 that is attached to the CSU marked DWEC448934.100."

# - Documentation and Verification

### Building a Logical Network Diagram

Documentation is critical to a successful network integration or cutover. Documentation can help you succeed or make you fail. There are several forms of documentation, and several ways of presenting it.

If the business you are auditing already has an information management system for the Help Desk / Network Control Center - you should update documentation in that application.

Desktop applications are another good way of presenting documentation. A spreadsheet can be used to maintain circuit IDs, cross-reference information, port information and costs. A word processing application can be used to maintain office contacts and notes. A graphics application can be used for drawing the network. This could be tied together by a database management program.

Microsoft Office is a good application suite for this purpose. You can use Microsoft Office to link Excel spreadsheets, Word documents and graphics together under PowerPoint. Microsoft's Internet Assistant simplifies the process of putting your documentation into HTML format. This lets you publish your network documentation on a Web server, so anyone on the company's Intranet can review the information.

There are also several network drawing applications available, that are designed specifically for logically depicting voice and data networks. NetViz from Quyen Systems is especially valuable in this area. NetViz contains graphic images of routers, hubs, PCs, Front End Processors, CSUs, and modems.

The application lets you create fields to contain information, and links between components. While NetViz is not designed for a large Enterprise network Help Desk, it will assist in documenting network connectivity, and lets you export your network map in HTML format. The graphics in this book were created by NetViz.

Graphics-based applications simplify network documentation and problem resolution by clearly indicating how network components are connected. If you see a router go down, it is helpful to know immediately what circuit it is connected to, and what vendor to call. If you are planning a network migration from one vendor to another, a graphic application can make your plans clear.

Whatever application you use, the important point is to maintain accurate documentation, and to make it available to anyone that needs it. Your documentation program could simply be a grease board that gets modified as network elements are moved, added or changed. Or you could invest in an expensive application suite that never gets updated. Whatever you use, just make sure it gets used.

### Documenting the Perceived Network

A perceived network is a network as presented by available information sources. This network consists of information elements and billing elements. The perceived network can be very different from the actual network. The task is to determine what the network looks like from a documentation and billing point of view, and compare it to the actual physical network components.

The physical components are also simply a piece of the network. The next step is to determine network connectivity - what physical elements connect to other physical elements, and what the network paths between the various application servers and clients are.

While building your documentation, you may start to notice anomalies. Some of the things you need to look for include:

Circuits that do not terminate in networking equipment.
Networking equipment that does not tie to a circuit.
Small offices with circuits from multiple carriers.
Small offices with multiple types of wide area networking
equipment.
Small offices with high circuit bandwidth.
Large offices with relatively slow speed circuits.
Networks with the same circuit speed at all remote offices.
Networks combining leased lines and frame relay for no
obvious reason.

The network documentation - as it appears from network billing detail and information management systems - is a good starting point in your audit. This is where you start to determine where circuits should be disconnected. This is where you can start to question circuit speeds in some offices. This is also where you can start to question the validity of using multiple carriers in smaller offices.

Circuits that do not obviously tie to networking equipment requires closer attention. Documentation may be incomplete, or you may have found a circuit that can be disconnected. This step will reduce your monthly bill.

If network documentation shows networking equipment that does not connect to a circuit, it can be due to faulty documentation, or to a network conversion.

A business may have moved from IBM SNA/SDLC equipment to multi-protocol routers. The business may have moved from a bridged environment to a routed environment. In either case, you need to determine if the equipment not in use was leased, if you are paying maintenance for it, and if its circuit was disconnected.

For an example, I recently saw invoices for ancient AT&T/Paradyne 2400 BPS modems on lease. The business was being charged $55 a month for each modem. The equipment was lost during an office move (several years ago). The office owed over 12 months of payments on the leased equipment. The amount owed would have purchased several modems.

The annual savings for just getting 3 modems off lease amounted to $1,980. This is a fairly common occurrence, so look at those leases!

Make sure you look at all lease agreements and can put your finger on the equipment being leased. You also need to note when the lease expires, to make sure the equipment comes off lease and payments do not turn to a monthly rental.

While you are verifying physical network hardware, look for maintenance agreements. In all too many cases, annual maintenance agreements keep getting purchased for equipment that has long since been disconnected. Do not sign maintenance contracts for equipment that you can not identify, as the equipment is probably sitting in some storage closet.

One last point about that network equipment that is not in use. Make sure to get an inventory of the equipment, its software level, and its configuration. Get the serial numbers as well. You may be able to sell the excess equipment, generating a small revenue for your business.

Prices for some equipment may be very small, but you should do well for older routers that the vendor still supports. You should also do well with smart hubs such as the Synoptics / Bay Networks and Cabletron families.

When you find offices with multiple carriers, you will need to determine why this has occurred. The office may have circuits from multiple vendors to provide backup in case one vendor has an outage. This is valid. You may find that the business migrated from one carrier to another, but missed a few circuit disconnect requests. If so, disconnect the excess circuits and keep track of your orders and cost savings!

As offices change in size due to business requirements, networks don't always keep pace. Small offices that expand often get increased bandwidth, as there is a great feedback mechanism when response times are poor.

This feedback mechanism is unsatisfied people! The network support personnel will hear, perhaps several times a day, that network response time is slow. This tends to get corrected as the level of pain starts to outweigh the cost of increased bandwidth.

As offices shrink in size, you may miss opportunities to decrease circuit speeds, or to decrease the frame relay CIR. This is because people tend not to call when the network is too fast. Your opportunity is to look for offices that are small in size and in networking requirements, but have as much capacity as larger offices. You may be able to decrease circuit speeds and reduce the monthly network bill.

A network discovery tool would be a great asset to this part of your project. There are tools available that will look at a network and show how many devices are on the network. There are tools for TCP/IP devices and for NetWare IPX/SPX devices. By using one of these applications you will know exactly how many devices are on each segment of a network.

Some discovery tools available for Windows PCs supporting TCP/IP and IPX/SPX networks include SNMPc from CastleRock and HP Openview for Windows. Unix tools offer a great deal more power, but may not be required for a basic audit of a relatively small to mid-size network.

While creating your documentation, you may find networks that combine frame relay and leased lines. In many cases it is perfectly fine to combine leased lines and frame relay. One reason would be if you have two large offices and several small offices. You could connect the large offices via a leased line, then use frame relay in the remote offices.

Each remote office could have a PVC pointed at both large offices. This would give you back up in the event that frame relay services went down at one headquarters office. Traffic could go through the frame relay port at your second location, then ride across the leased line. This would give you some back up.

As a network auditor, you should question it if you see an office with a leased line when frame relay would be able to handle the traffic. You may have run across an office that was brought up before the business started using frame relay services.

The office may just have never gotten converted to the new network structure. A quick look at leased line costs versus frame relay costs may help you decide if the office should be migrated, or if the office should stay on a leased line.

## Verifying Multi-Protocol Network Connectivity

How do you verify connectivity in a multi-protocol network? Glad you asked! In your research, you should have found circuit IDs and a general idea of hardware in place. If you went on a site survey, you should have been able to find physical circuits and identify what network equipment they are connected to.

After verifying physical connectivity, you can also verify logical connectivity. In our samples we will use some tools at hand to connect to a Cisco router and verify interfaces and frame relay mapping.

You can connect to a Cisco router (or any other multi-protocol router) in one of two ways. You can either dial into a modem attached to the router, or you can telnet to the router. In some cases IP filters may be in place to block access. There may be network equipment in place, but you may not be able to connect to it through telnet. If you can't connect via telnet, then try to dial into the modem port.

After connecting to the router, you will need to enter your password. You will want to check the interfaces on the router. You may not need to verify protocols, IP addresses, router memory and software levels for a financial audit; but you should gather this information for a network audit.

First we will check the status of interfaces, and verify what protocols are running on a Cisco router. To do this, connect to your router. Enter your password. At the command line type in "sh proto". This will show running protocols on the router.

## Show protocols - verify port status, and check protocols

```
CUSTRTR > sh proto
Global values:
Internet Protocol routing is enabled
Novell routing is enabled
Ethernet0 is up, line protocol is up
  Internet address is 3.79.236.1, subnet mask is 255.255.252.0
  Novell address is 34FEC00.0000.0c3e.a5dd
Serial0 is up, line protocol is up
  Internet address is 3.119.124.27, subnet mask is 255.255.252.0
  Novell address is 3777C00.4000.0379.2361
Serial1 is up, line protocol is up
  Internet address is 3.155.88.20, subnet mask is 255.255.252.0
  Novell address is 39B5800.4000.0379.2361
```

This tells us that the router is running TCP/IP protocols (Internet routing) and IPX/SPX protocols for Novell routing. This also tells us that the router has one Ethernet port connecting it to the Local Area Network. The router has two serial interfaces connecting it to the Wide Area Network. The Ethernet port, and both of the serial ports are up and connected (Serial 0 is up, Line protocol is up). This may be of concern, so we will check why a small 2500 router with two serial ports has both ports up and active.

The next thing to verify is the software level and amount of memory that is in the router. This will not effect a financial audit, but is essential for network planning and design. To get this information, enter "sh ver" from the command line interface.

Copyright (c) 1986-1994 by cisco Systems, Inc.
Compiled Tue 25-Oct-94 19:13 by dougs

ROM: System Bootstrap, Version 5.2(5), RELEASE SOFTWARE
ROM: 3000 Bootstrap Software (IGS-RXBOOT), Version 10.2(5), RELEASE SOFTWARE (fc 1)

CUSTRTR uptime is 10 weeks, 6 days, 5 hours, 3 minutes
System restarted by power-on
System image file is "flash:igs-bpx-l.100-6", booted via flash
Host configuration file is "custrtr", booted via tftp from 3.61.248.200
Network configuration file is "pcmdbAAAa1787113", booted via tftp from 3.61.248.200

cisco 2500 (68030) processor (revision D) with 16380K/2048K bytes of memory.
Processor board serial number 02092907
X.25 software, Version 2.0, NET2, BFE and GOSIP compliant.
Bridging software.
SuperLAT software (copyright 1990 by Meridian Technology Corp).
Authorized for Enterprise software set.  (0x0)
1 Ethernet/IEEE 802.3 interface.
2 Serial network interfaces.
32K bytes of non-volatile configuration memory.
8192K bytes of processor board System flash (Read ONLY)
Configuration register is 0x2102

This tells us that CUSTRTR is running 10.2(5) software, and it has 16 meg of memory. This also lets us know what protocols the router is licensed to support. You will need to know this to make sure the router can support newer releases of software.

Now that we have an idea of router interfaces, protocols, network addresses, software levels, and memory - let's learn a little more for the network audit!

You can check the status of an interface by entering "sh int" followed by the interface number at the command prompt. We are interested in the serial interfaces as these are connected to circuits that appear on your monthly network bill. To check on serial interface 1, enter "sh int s 1".

```
CUSTRTR1 > sh int s 1
Serial1 is up, line protocol is up
  Hardware is HD64570
  Description: S1: CUSTRTR frame link 128kb  DLCI #117
  Internet address is 3.155.88.20, subnet mask is 255.255.252.0
  MTU 1500 bytes, BW 128 Kbit, DLY 20000 usec, rely 255/255, load 1/255
  Encapsulation FRAME-RELAY, loopback not set, keepalive set (8 sec)
  LMI enq sent  100003, LMI stat recvd 74853, LMI upd recvd 4, DTE LMI up
  LMI enq recvd 0, LMI stat sent  0, LMI upd sent  0
  LMI DLCI 1023  LMI type is CISCO  frame relay DTE
  Broadcast queue 0/64, broadcasts sent/dropped 135055/0
  Last input 0:00:00, output 0:00:03, output hang never
  Last clearing of "show interface" counters never
  Output queue 0/40, 0 drops; input queue 0/75, 0 drops
  Five minute input rate 1000 bits/sec, 2 packets/sec
  Five minute output rate 0 bits/sec, 0 packets/sec
     237307 packets input, 44293530 bytes, 0 no buffer
     Received 0 broadcasts, 0 runts, 0 giants
     3 input errors, 3 CRC, 0 frame, 0 overrun, 0 ignored, 3 abort
     385670 packets output, 66970521 bytes, 0 underruns
     0 output errors, 0 collisions, 163252 interface resets, 0 restarts
     23 carrier transitions
     DCD=up  DSR=up  DTR=up  RTS=up  CTS=up
```

This tells us that a circuit is connected to serial interface 1. It also tells us that the circuit is good! Serial 1 is up / line protocol is up tells us that the circuit is good, and something is connected on the other end. We can also tell that this is a frame relay circuit, and that the circuit is seeing LMI from the carrier. That means that we are good to the vendors frame cloud. We can tell this by seeing LMI Up status.

We may want to keep an eye on circuit utilization, as this router has a 128KB circuit and averaged 1000 bits per second during the last 5 minutes. The frame relay provider can provide you with network statistics that help clarify utilization. Alternatively, a good Network Management system should also give utilization statistics.

Following is an example of a "sh interface" command for a serial port that should be up, but isn't up yet. If the circuit associated with this interface appears on the bills, it should be questioned. The circuit should not be accepted for billing purposes until it has been tested and is up and passing data.

```
CUSTRTR2 > sh int s 0

Serial 0 is down, line protocol is down
  Hardware is MK5025
  Description: S0: CUSTS MCI frame link 128  DLCI #111
  Internet address is 3.155.88.14, subnet mask is 255.255.252.0
  MTU 1500 bytes, BW 128 Kbit, DLY 20000 usec, rely 255/255, load 1/255
  Encapsulation FRAME-RELAY, loopback not set, keepalive set (8 sec)
  LMI DLCI 1023, LMI sent 173162, LMI stat recvd 173162, LMI upd recvd 0, LMI N3
91 6
  Last input never, output never, output hang never
  Last clearing of "show interface" counters never
  Output queue 0/40, 0 drops; input queue 0/75, 0 drops
  Five minute input rate 0 bits/sec, 0 packets/sec
  Five minute output rate 0 bits/sec, 0 packets/sec
    0 packets input, 0 bytes, 0 no buffer
    Received 0 broadcasts, 0 runts, 0 giants
    0 input errors, 0 CRC, 0 frame, 0 overrun, 0 ignored, 0 abort
    0 packets output, 0 bytes, 0 underruns
    0 output errors, 0 collisions, 46179 interface resets, 0 restarts
    46187 carrier transitions
```

In this case, the circuit had not been installed yet. A circuit could be down/down for three reasons:
1.  It has not been dropped off by the Telco
2.  The CSU has not been installed
3.  The CSU has been installed, but is not cabled to the router

If Telco is late, the circuit should not be billed. If the CSU has not been installed because the customer did not extend the Dmarc, the circuit can be billed but will not work. If the router is simply not connected to the circuit, connect that V.35 cable!

In this case, Telco was late. The customer was billed for 6 weeks of service that was not received. The carrier issues a credit in this instance.

This is an example of a T1 circuit that is installed, connected to a CSU and should be active.

```
CUSTRTR > sh int s3/5
Serial3/5 is up, line protocol is down
  Hardware is cxBus Serial
  Description: S3/5: Link To BCKUPRTR
  Internet address is 3.107.8.1 255.255.252.0
  MTU 1500 bytes, BW 1544 Kbit, DLY 20000 usec, rely 255/255, load 1/255
  Encapsulation HDLC, loopback not set, keepalive set (8 sec)
  Last input never, output 0:00:06, output hang never
  Last clearing of "show interface" counters 0:07:11
  Output queue 0/40, 0 drops; input queue 0/75, 0 drops
  5 minute input rate 0 bits/sec, 0 packets/sec
  5 minute output rate 0 bits/sec, 0 packets/sec
    0 packets input, 0 bytes, 0 no buffer
    Received 0 broadcasts, 0 runts, 0 giants
    0 input errors, 0 CRC, 0 frame, 0 overrun, 0 ignored, 0 abort
    54 packets output, 1296 bytes, 0 underruns
    0 output errors, 0 collisions, 15 interface resets, 0 restarts
    0 output buffer failures, 0 output buffers swapped out
    0 carrier transitions
    RTS up, CTS up, DTR up, DCD up, DSR up
```

From this we can see that a circuit is physically connected to the port, and that the circuit is up and running. We know this by seeing that the port is up, and that it sees Request To Send, Clear To Send, Data Transmit Ready, Data Carrier Detect, and Data Set Ready.

We can also know that nothing is connected to the CSU on the other end of the circuit. We can tell that by seeing line protocol down.

If this is a backup circuit, we would need to make sure that the circuit is tested, and that the end user knows that manual intervention is required. The CSU will need to be connected to a router on the far end, possibly the router will need to be re-configured.

This could also point out a circuit that should be disconnected.

The next thing to do, is to check what routers this router connects to. As frame relay is not a point-to-point network, we do this by checking frame mapping and frame PVCs. Frame mapping lets us know what the router sees on the other end. This can be dynamic, or statically defined. Following is an example of the "sh frame map" command:

```
Serial1 (up): novell 39B5800.0000.30c0.79da dlci 30(0X1E,0X4E0), static, broadcast,
        CISCO, status defined, active
Serial1 (up): novell 39B5800.0000.3061.2521 dlci 40(0X28,0X880), static, broadcast,
        CISCO, status defined, active
Serial1 (up): ip 3.155.88.1 dlci 40(0X28,0X880), static, broadcast,
        CISCO, status defined, active
Serial1 (up): ip 3.155.88.2 dlci 30(0X1E,0X4E0), static, broadcast,
        CISCO, status defined, active
```

From this we learn that CUSTRTR is using TCP/IP and Novell's IPX/SPX protocols over the wide area network. We can also see that the router is mapped to DLCIs 30 and 40, and that the DLCI status is active. As carriers normally bill for DLCIs or PVCs, it's important to verify that you can see them, and that they are active!

This is a very simple frame map. A central router serving multiple frame clients will have more information.

The next step is to verify that we are actually sending and receiving data over each PVC. We can verify the frame relay configuration of serial port 1 by entering "sh frame pvc int s 1". Following is sample output from the sh frame pvc command:

```
CUSTRTR > sh frame pvc int s 1

PVC Statistics for interface Serial1 (Frame Relay DTE)

DLCI = 30, DLCI USAGE = LOCAL, PVC STATUS = ACTIVE, INTERFACE = Serial1

    input pkts 36059      output pkts 84690      in bytes 7010620
    out bytes 29729282    dropped pkts 5         in FECN pkts 0
    in BECN pkts 0        out FECN pkts 0        out BECN pkts 0
    in DE pkts 36059      out DE pkts 0
    pvc create time 1w6d  last time pvc status changed 13:04:45

DLCI = 40, DLCI USAGE = LOCAL, PVC STATUS = ACTIVE, INTERFACE = Serial1

    input pkts 68703      output pkts 69869      in bytes 22204228
    out bytes 27921769    dropped pkts 18        in FECN pkts 0
    in BECN pkts 0        out FECN pkts 0        out BECN pkts 0
    in DE pkts 68703      out DE pkts 0
    pvc create time 1w0d  last time pvc status changed 11:00:33
```

The information above tells us that we see PVCs 30 and 40, and that the PVCs are active. We also know that we are sending and receiving data over both PVCs. Finally - CIR sizing is not too small, as we are not seeing BECN or FECN packets. We did drop a few packets, but the application retransmits further up the OSI model. This is running very clean.

## Verifying SNA/SDLC Network Connectivity

SNA Connectivity is a fairly simple thing to verify. A remote control unit (an IBM 3x74, or a device that emulates a 3x74 PU Type 2) connects to the network through a CSU and a Front End Processor.

These devices are identified to the network through an NCP gen. This is software that resides in an IBM 37X5 Front End Processor. An IBM Control Unit (PU Type 2) connects to a CSU, and this connects to a circuit. The circuit connects to an IBM Front End Processor.

You can check the status of the circuit by displaying the line in Netview. The circuit will be logically associated with a line. You can check the line by entering "d net, id=(line name)". If the line shows active, then the logical line is associated with a physical circuit and the physical circuit is up and running. If the line shows as inactive, then there is a question with the line.

The line connects to a control unit. To make sure the control unit is still connected and powered up, type in "d net,id=(controller name)". If the controller shows as active, it is up and connected. If the controller shows as inactive, it may have been inactivated in the NCP gen (coded istatus-inactive), or it may have been manually inactivated. If the controller shows as PCTD2, then the line is up an the controller is down. In this case the controller will have been coded as istatus=active.

After verifying line and controller status, you should check the status of attached terminals and printers. There have been numerous occasions when circuits where up (and appearing on the monthly bills), and control units have been up (and appearing on maintenance contracts) when all the terminals and printers have been disconnected and the office moved to a LAN environment. If an office has multiple IBM control units, you can check each controller to see if terminals and printers can be consolidated. This can give you the opportunity to remove some networking equipment and reduce your maintenance costs.

## Shutting Down Unused Resources

If you identify equipment that has been coded as active, but in fact should be inactive, you should change the status to inactive. For a Cisco router, you go into enable mode and issue a shutdown command for the appropriate interface. For IBM SNA equipment, you would issues a "v net, inact,id = (line name)" or "v net, inact,id = (controller name)". Each hardware vendor will have their own command for shutting down a port.

By inactivating equipment interfaces through a network management application, you will have a quick fall back plan in case the equipment port is in use. You will simply be able to go back out and activate the port.

It is critical to inactivate resources before issuing circuit disconnects. As new circuits are dependent on the telco and can take 30 days or longer, it's better to be absolutely sure equipment is not in use before issuing the disconnect request.


## Disconnecting Unused Resources

After inactivating resources that appear to not be in use, you should keep track for at least two weeks. You will be able to make sure that the circuits are not in use.

First check to see if information is flowing over the circuit. Then manually inactivate the circuit. If there are no complaints, issue a disconnect. If there are complaints, manually activate the network equipment and make note of the user that requested it to be activated. Also note the application this network equipment serves.

When placing disconnect requests, make sure to keep a copy of all orders. Keep your request information, the day the request was placed, the person you notified (in writing) and their work order number.

It often takes 30 days or more for circuits to be disconnected, so you also have a back-out plan in case the circuits you disconnected are required for a new project.

**Verify those Disconnects!**

How do you verify those disconnects actually took place? While placing disconnect requests, always keep a copy of your order. Also make sure to keep a copy of the vendors work order number.

As you review your monthly bill statements, make sure that you are no longer being billed for circuits that you placed disconnects on. If the circuits are still on your bill, you can request a refund by referring to your original disconnect request and the vendors work order number.

You can also check disconnects by trying to activate resources. If you placed a request to disconnect a frame relay PVC, check PVC status and see if you can see it. If you placed a request to disconnect a leased line, you can try to bring the circuit up. If it comes up more than 30 days after you placed the disconnect, you should notify the vendor.

# Part III - Putting It Together

## - The Many Faces of Documentation

### Documentation for the Help Desk

The Help Desk will require documentation that will help them identify the remote office, and provide first level problem determination information. This includes addresses, names, and phone numbers for remote offices. This should be verified at least once a year, as offices move or people change positions. Help Desks are normally not notified, and documentation is rarely updated.

As a consultant, part of your project should be to verify the accuracy of the Help Desk's information, and to update it. A Help Desk should not have to spend time looking up the correct contact in the event of a network outage.

All contact names, phone numbers and pager numbers should be readily available and accurate.

### Documentation for the Network Control Center

The Network Control Center will require a greater level of documentation than the Help Desk. Information required for the Network Control Center will include Wide Area Network equipment information. This includes equipment type, equipment manufacturer, and equipment models.

The network control center will also need to know what interfaces are in use, what protocols are supported, and what TCP/IP or IPX/SPX addresses are assigned. The network control center will need to know how the remote equipment communicates over the wide area network. This will include circuit IDs, DLCIs (if Frame Relay), CSU vendor, and T1 Multiplexor vendor.

The network control center will need to know who maintains the CSU and/or T1 multiplexor. In addition to circuit connectivity, the control center will have to know who to call if a circuit is down, and how the call gets escalated in the event that the remote office experiences a significant outage.

While it is important to know what networking equipment is on site, the control center will also need to know if there is a maintenance agreement in place, and who to call if there is an equipment failure. The control center will also need to know if there is a modem connected to the networking equipment for remote maintenance.

If the remote office has a support person that is on call, the control center will need that person's office phone number and pager number.

If a router is down, you do not want to have the Help Desk trying to determine the correct contact, and then have the Network Control Center call the wrong circuit in.

While this seems farfetched in a small network, imagine a center with responsibility for several hundred routers, and hundreds or thousands of circuits.

If the business being supported has had an active history of moving from one vendor to another, it would be very easy for the information database to have the wrong contact, the wrong address (we moved offices within the same area and kept the same phone numbers and circuit IDs several times), and the wrong circuit ID.

All of this needs to be verified at least once a year, and possibly every quarter. The time spent in verifying information can pay for itself in decreased downtime during an emergency.


### Documentation for the Remote Office

Just as the Help Desk and Network Control Center need good documentation, the remote office needs documentation as well. The remote office will need to know who to call when there is a network failure.

The remote office will need to know how to identify themselves to the help desk. In a small network, office location may be sufficient. In a large network, office IDs or office numbers may be assigned.

Just as the Network Control Center needs circuit IDs and network information, the remote office needs the same. Each circuit should be labeled. Each CSU should be labeled with the carrier's name and the circuit ID. Every cable attached to a router or bridge should be labeled.

The router or bridge should have its network name written on it. The router or bridge should also have the circuit ID labeled on it, so the office contact won't have to trace a cable back to a CSU.

The router or bridge should also have the Help Desk's phone number written on it. If the network goes down, you don't want the office contact looking through files for the right person to call.

Each office should have a Service Level Agreement, and a contact list. The contact list should have an escalation list on it as well. The escalation list could state that if the network goes down, call the Help Desk first. If that doesn't resolve the problem, contact the Help Desk shift supervisor.

If the problem persists, the next call would be directly to the Network Control Center, then to the Network Control Center manager. If this is a commercial account, calls should also be made to the account representative.


### Documentation for Budgeting
Network managers need to be able to budget for network growth or network consolidation. The network planner or network manager will need to know what equipment is in place. The planner will need to know how much memory is in the equipment, and what software level the equipment is running on.

The planner will need to know what circuits are in use, what the monthly payments are, and what the incremental cost is to increase circuit speed, port speed, CIR, or to add PVCs pointing to another location.

As the business shares its plans for office growth in one area, and possible office consolidation in another, the planner can take this into account. As the Information Systems group shares plans for the development of distributed client-server applications to replace or enhance telnet or TN3270 based applications, the planner can start to budget for increased circuit speeds.

If the business is heading towards an Intranet structure with graphics based applications such as Web Servers, the planner will need to make sure there is network capacity to handle the increased traffic loads.

A good starting point would be to trend wide area network utilization. By trending utilization, you will be able to see when you may start to require additional bandwidth. You will also be able to see when you can reduce bandwidth.

By determining your utilization trends over a period of time, and determining the incremental costs associated with moving from a 128K port to a 256K or higher speed port, you will be able to budge for growth.

The key point in budgeting for growth is that a 56K frame relay port comes in over a 56K circuit. Fractional T1 frame relay ports come in over a T1 circuit. If you already have FracT with a 128K circuit, the incremental is relatively minimal to increase to 256K, 384K, 512K, ...

If you have a 56K circuit and require an increase, the costs are substantial as you will require a new access loop and a new CSU.

# - Publishing Cost Savings

As the network audit comes to a close, you will have the opportunity to publish cost savings. Information that should be published includes:

Circuit disconnects

Canceled maintenance contracts on unused equipment

Equipment taken off lease

Cost savings associated with lower speed circuits

Modem pool analyses

Unused equipment sold to secondary market

Savings should be noted by type, and posted as monthly and annualized savings. A monthly savings of 2K doesn't sound like much. An annualized savings of 24K is a completely different animal.

# Part IV - Right-Pricing the Network
# Network Optimization

## - Putting the Network Out to Bid

**Data Network Only, or Combined Voice and Data?**

As you move forward in right-pricing the network, you will have to make some hard decisions based on cost and performance. By using integrated access, you can share a T1 circuit between both voice and data.

The voice portion can include Long Distance outbound, Virtual Private Network for both inbound and outbound calls, non-channelized 800 traffic inbound, and data communications in the form of leased lines or frame relay.

As both voice and data can ride over the same T carrier, you can reduce your monthly bill by sharing access on the local loop. You can also save on T1 multiplexor equipment by using one port on your mux for voice, and a second port on your mux for data.

This can be an elegant solution, but does have some drawbacks. These may include:

> **Disadvantages:**
> By using one carrier for both voice and data, you can make future network conversions more complex.
>
> If the T1 goes down, your office will lose its data communications. It will also lose a large portion of voice capacity.
>
> You will need to be aware of both voice and data growth trends, to be able to meet growing needs.
>
> **Advantages:**
> Reduced monthly voice and data communications bills through integrated access.

Lower cost per minute for calls over dedicated facilities

Reduced need for local lines as LD will go over dedicated facilities.

Another advantage is even further reducing network costs by bidding both the voice and data portion, and leveraging your combined volume.

The potential cost-savings may outweigh any objections. Especially if you already have a plan to move 800 calls over to your DID service if your T1 goes down, and if you have ISDN BRI as a backup plan for your data communications network.

As this can easily be the case, let's look at planning for integrated access. The information you need includes:

**Data**
Port speed required for current applications
Estimate of short term requirements
Estimate of long-term requirements

**Voice**
PBX/Hybrid manufacturer
PBX/Hybrid model
Cost to add T1 card (or channel bank)

**Telco**
Installation charges for T1
T1 monthly re-occuring local loop charges

Average minutes per month of LD traffic
Average minutes per month of on-net traffic
Average minutes per month of International traffic

Number of channels required to support voice traffic
Cost per co-trunk
Number of co-trunks that could be off-set

This can be further clarified by the example on the following page.

# Break-Even Analyses - T1 versus Switched Access

| One Time Costs | | Re-Occuring Charges | |
|---|---|---|---|
| **T1 Analyses** | | **Cost Take Out (Remove CO Trunks)** | |
| Price of T1 card | 1,600.00 | Cost for CO Trunk | 43.75 |
| Price of T1 CSU | 0.00 | Number of trunks replaced | 12 |
| Software Costs | 0.00 | | |
| Installation charges (card) | 300.00 | Reduction in local bill | 525.00 |
| Switch upgrade charges | 0.00 | | |
| Telco T1 Inst. Charge | 680.00 | | |
| *Investment* | 2,580.00 | Average minutes per month LD/On-net | 38,526 |
| | | Cost per minute: switched access | 0.103 |
| | | Cost per minute: dedicated access | 0.070 |
| | | Reduction - Dedicated versus switched | 1,271.36 |
| *Months to pay back* | 2.52 | Increased costs: T1 loop charge: | 1,023.68 |
| *Years to payback* | 0.21 | Net change in monthly bill: | 772.68 |

This tells us that we will spend $1,600 for a T1 card. Looks like we're going to the secondary market! We are not including the CSU charge, as one is required for the data communications network anyway. We are estimating that installation of the T1 card will run $300.

The office currently has about 12 channels worth of Long Distance and On-Net calling traffic. By moving the calls over to a shared T1, we will remove 12 CO-trunks. We are currently paying 43.75 per co-trunk, so our local bill will be reduced by $525.00 per month. We should use a good call accounting program to verify trunk requirements.

Alternatively, we could use an application like Lucent Technologies WINPBX IMS and TGA (Trunk Group Analyzer) if we were working on an AT&T/Lucent PBX. Nortel and Siemens/Rolm have similar tools. In this case, we just cowboy'd the requirements by reviewing Harry Green's "Telecommunications Management" book, and cross-referencing "The

Complete Traffic Engineering Handbook" by Harder, Wand and Richards, Jr. By reviewing both books, we came up with the following:

Average of 38,526 minutes per month billed LD, International and 800 inbound
Average of 20 working days per month of billed traffic
Office has 8 hour working day

| 38,526 / 20 | = 1,926.3 average call minutes per day |
| 1,926.3 * .17 | = 327.471 average call minutes during busy hour |
| 327.1 * 60 | = 19,648.26 call Seconds |
| 19,648.26 /3600 | = 5.45785 Erlangs |

5.45785 Erlangs requires 12 channels to provide a P.01 grade of service

We determined the average working days per month by interviewing office staff and looking at the calendar. The office is closed on weekends.

We determined the average working day by interviewing office staff.

The office currently pays 10.3 cents per minute for switched access long distance calls. As the current office is off-net, we are estimating on-net calling patterns (but should have a good idea by looking at other offices of similar size and patterns). By moving to dedicated access, we will reduce our per minute rate to 7.0 cents per minute. The difference between 10.3CPM and 7.0CPM for our average 38,526 minutes per month brings us another $1,271.36 per month savings.

We included local loop charges in this example. With integrated access, the local loop charge would be shared for both voice and data. We should be able to improve on the savings noted above. As local loop charges are mileage dependent, a loop will normally run between $580 and $1,300 dollars per month. By sharing the loop between voice and data, we are able to off-set charges for one T1. This amounts to a $580 to $1,300 per month savings, or an annualized savings of $6,960 to $15,600.

In this example we were able to reduce our monthly voice bill by removing co-trunks and achieving a lower LD rate. We increased costs by paying for a T1 local loop. The amount we reduced the monthly voice portion of the telecommunications bill was $772.68. This is an annualized $9,272.16. We also reduced the telecommunications bill by sharing the loop between voice and data. As the loop in this case costs $1,023.68 and we used one, not two - we were able to further offset the monthly re-occurring charges by $12,284.16. Our total savings are an annualized $21,556.32.

This model can change in several ways: T1 access costs can vary significantly. CO-trunk costs can vary. Differences in rates for switched access versus dedicated access can vary. PBXs or Hybrids may need an additional cabinet to support a new T1 card, greatly increasing the costs to add T1 support. In addition, the number of trunks required to carry a given level of traffic will vary depending on your specified grade of service.

While it is best to use a call-accounting package or raw data from the PBX, you can make some estimates based on a given number of minutes of traffic, the number of working days in the billed month, the length of the working day, and a specified level of blocking. By tracking Long Distance volume by office in a spreadsheet, you will start to notice when offices are candidates for T1.

Following is a diagram of integrated access. While potential savings look good in one instance as noted above, this can easily scale across the network, multiplying savings by the number of offices targeted for integrated access.

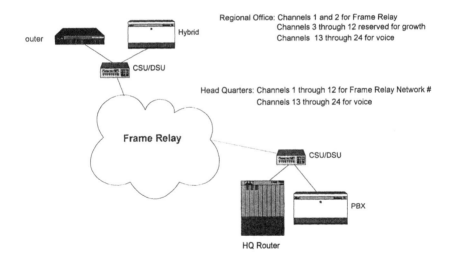

Each office has T1 Circuit to carrier's network.
Voice is muxed out on DACS at carriers Point Of Presence (POP).
Disaster Recovery: 800 Numbers inbound overflow to DID service
ISDN BRI used to back up data communications.

**Migrating from Leased Lines to Frame Relay**
Christine Heckart's "The Guide to Frame Relay Networking" from Flatiron Publishing is an excellent reference on frame relay networking, frame relay devices, and selecting carriers. If you are interested in frame relay, just get the book. It is one of the best around.

Why should you consider migrating from a leased line environment to a frame relay environment? You would migrate to reduce network costs substantially. You would migrate to increase connectivity at minimal costs. You would migrate to reduce access costs, as frame relay allows a one-to-many configuration.

Frame Relay carriers have also been known to oversell in a big way. Your best defense against being oversold is to read Christine's book and follow the guidelines.


**Optimizing Frame Relay**
There are several tools available for optimizing frame relay networks. The points to be aware of include:

      Circuits can be 56K or T1

      Port speeds can be:
          56K, 64K, 128K, 256K, 384K, 768K, 1024K, or 1536K

      CIR can be:
          0K, 4K, 16K, 32K, 48K, 56K, Nx64K

      PVCs are billed items

As management reports from the frame relay provider can be available at no charge to very little charge, and the reports cover billed items - they are a perfect place to start. We will be able to verify port speeds requirements, CIR requirements, and PVCs.

We will not be able to verify mileage dependent access charges with a carrier's report, and may want to partner with a tariff specialist.

MCI offers Hyperstream reports via a Web server and via FTP. The other carriers offer reports as well. If you are performing network audits and network optimization as a consultant, the reports should be billed to the client.

Reports should be trended for 30-45 days to get a good idea of average utilization. In trending reports, look at how each item correlates to the bill.

Frame Relay lets you have a specific port speed and a committed information rate (CIR). The CIR is the minimum amount of information in terms of bits per second that the carrier will guarantee. Above this, you may experience dropped packets.

The port speed is your maximum circuit speed. When the network is not experiencing congestion, you may be able to burst up to the full port speed.

A PVC is a private virtual circuit mapping one location to another. At the very least, each of your remote offices will need a PVC back to your headquarters office. Above this, you may map PVCs between offices if you have distributed applications and non-centralized servers.

How do you optimize a frame relay network? By looking at average utilization and peak utilization. Some rules of thumb are:

Port Speed Sizing:

      If utilization is consistently less than 20%, decrease port speed

      If utilization is consistently above 60%, budget for increased port speed

      If utilization is consistently above 80%, place orders NOW!

CIR Sizing:

      If you are experiencing a large number of dropped packets, increase CIR

      If you are running at 98-100% throughput, CIR can be reduced

PVCs:

> If you are not sending much information over a PVC, and if
> you have an alternate path , consider removing a PVC.

The following pages are based on graphs created using vendor provided management reports. The information was available at minimal cost and effort. The time required to create the graphs was minimal. The savings potential more than pays for the time and effort required in creating and formatting the reports.

In this example we are checking serial port utilization on a frame relay network. As a serial port is full duplex, it can send and receive in bursts up to port speed. As a refresher, we are looking for utilization consistently below 20% as a potential for reducing port speed. We are also looking for utilization above 60% to budget for future increases, or above 80% for immediate action.

The report gives us percentage utilization of the port in both directions. The report also tells us how much data was transmitted and received. We can look at the dropped bits section, to determine percentage of bits dropped that had the DE (discard eligible) bit on, versus non-DE packets.

## Serving Site A T1 - Frame Utilization

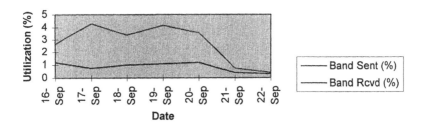

## Client A 128KB - Frame Utilization

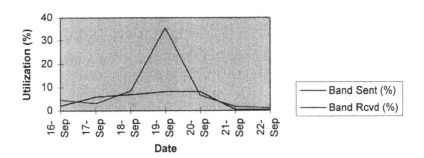

## Client B 128KB - Frame Utilization

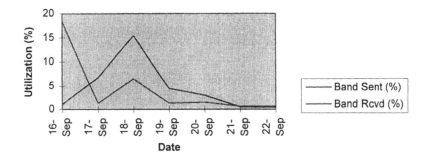

## Client C 128KB - Frame Utilization

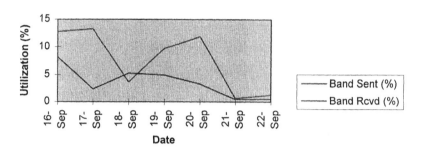

## Client D 128KB - Frame Utilization

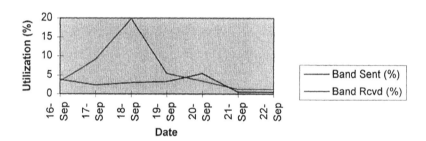

## Client E 128KB - Frame Utilization

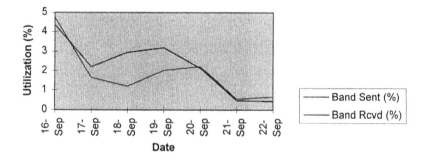

# Client F Frame Utilization

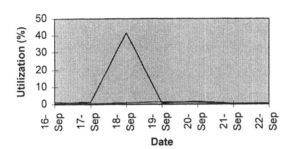

# Client G  Frame Utilization

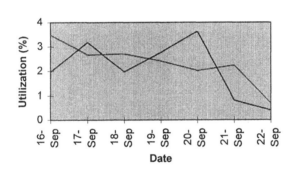

# Serving Site B T1 - Frame Utilization

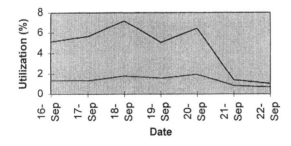

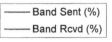

## Test Site 128KB - Frame Utilization

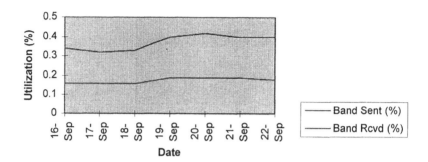

## Client H 128KB - Frame Utilization

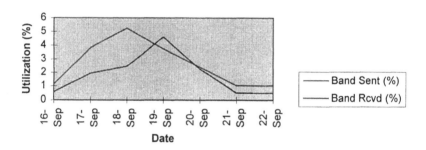

## Client I 128KB - Frame Utilization

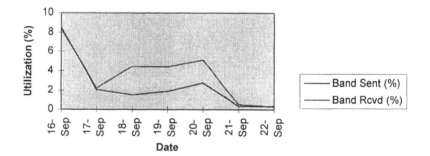

# Client J 128KB - Frame Utilization

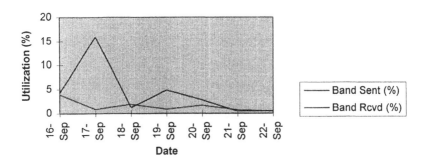

# Client K 128B - Frame Utilization

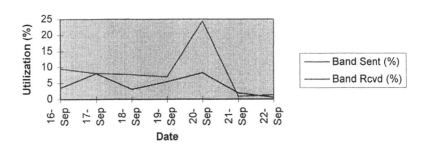

# Client L 128K - Frame Utilization

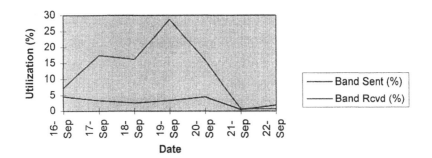

## Client M 128KB Frame Relay Utlization

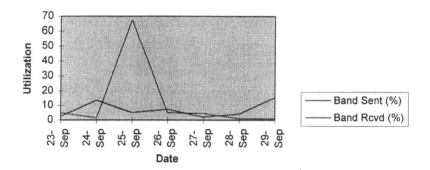

## Client N 128KB - Frame Utilization

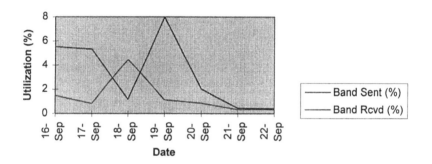

While we are only looking at one weeks worth of information, we can make some generalizations about the client's network. By reviewing utilization for a month to 45 days, we will have enough information to right-size the network.

Following is the working fact set:
>All remotes are communicating over 128KB ports
>Each remote has a 0K CIR
>A separate report shows throughput.
>>Each remote is running at 98%-100% throughput
>Each remote has a PVC to serving sites A and B
>Serving sites have full T1 ports

Network is migrating from character-based applications running over Telnet and TN3270 to client-server applications and graphics based applications such as Web sites

As an auditor working on a percentage of savings, the recommendation may be to reduce all offices averaging less than 20% utilization to a 56K port. The recommendation might also be to reduce the serving sites from 1,536KBPS (full T1 port speed) to 768KB. Estimated savings are 55 to 65K annualized.

As the network is evolving, and as 56K circuits come in over different facilities than FracT - it may make more sense to keep the majority of the offices on 128KB, even though they are underutilized.

The rational for this is:

To reduce from 128K, 56K circuits will need to be ordered

After the sites have been converted to 56K, disconnects will need to be placed on the 128K circuits

After the client-server applications go into production, speeds may need to be increased. This would require new circuits to be ordered, and disconnects to be placed on the 56K circuits.

Therefore:

We will keep the majority of the offices on 128K as the minimum speed for Fractional T1 access.

We will move low utilization offices (less than 10%) utilization to 56K circuits.

We will move extremely low utilization offices to ISDN BRI dial-on-demand routing.

We will continue monitoring utilization as client-server applications are rolled out. If utilization does not increase to the target rate, we will reduce port speeds.

As we move forward, we would continue trending utilization to see when we should increase speeds, but it looks like there's a lot of capacity not being used.

Lastly, we will look at routing protocol choices as the user population is complaining about network response times. We can clearly show that it is not a utilization issue.

We also know it is not a case of dropped packets, as throughput is high. We may have a problem with SAP broadcasts every 60 seconds by using RIP or IGRP routing protocols. We may also have a problem with IP route broadcasts. The solution could be in going to EIGRP or OSPF. The solution may also be in tuning the applications. Bandwidth will not solve the problem, as we aren't using what we have now.

### Include Network Documentation
When putting the network out to bid, you will need to include all the network documentation that was previously gathered. This will include every office location, and all port speeds. This will also include CIR requirements, and documentation on PVC mapping.

If you decide to combine voice and data, you will also need to include the average minutes per month for each location. This should be broken down by long distance, on-net and International calls.

### Determining Growth Plans
Growth is essential to a company's health. As a company grows, its voice, data and video communications networks will need to grow as well.

Growth can be in the form of additional people. This could be new offices opening up, or an increased headcount. This could also be due to an acquisition.

Growth can also be in the form of increased productivity. The business may not necessarily hire more people, but the business may require an increase in productivity.

One way the network planner can determine growth and budget for network expansion is to meet with representatives from each department. If a department has an increased headcount, there will be an increased demand for network services.

If a department expects to open a new office, or acquire another business, there will be a greater demand for services. A business may also plan to consolidate offices. This will lead to a lesser demand for physical circuits, but a greater demand for increased circuit speeds.

The network planner may also be faced with an increase in client-server applications. Even if the number of networked devices is not increased, the number of actual bits being transmitted can increase greatly.

The network planner will need to make applications groups aware of the amount of time required to increase network speeds. The network group should be involved as applications are developed, to make sure the network can support the application.

Networks are built to support applications, not the other way around. Applications are written to serve a business's information needs. The driving force in network growth has to be an increased demand for information. This demand has to meet a business's goals of improving financial performance.

By being closely involved with Information Systems management, the network group can learn of applications before they become a reality. By being closely involved with LAN support and administration, the network group will know when additional communicating client devices are being deployed. By trending increased voice traffic, the network group will be able to project increased data communications network traffic.


**Qualifying Vendors**
Before including a vendor in an RFP or RFQ process, the vendor must first pass qualification. The following items can help qualify a vendor:

> Vendor Name
> Name of Account Executive
> Length of time in business as carrier
> Length of time as network provider
> Copy of Dun and Bradstreet report
> Annual report


The next sections, on Request For Information, Request For Quotation and Request For Proposal can help in qualifying vendors. Before sending an RFP out to vendors, it is beneficial to qualify them to some extent. The vendor's response to an RFP will help to qualify or disqualify the vendor.

---

## - RFIs, RFQs and RFPs

### Creating the RFI

An RFI is a Request For Information. It is used as a formal means for requesting information from a vendor. As network offerings are constantly changing, the RFI is a good way of keeping current with vendor offerings.

A Request For Information does not obligate the client to purchase anything, although it does signify an interest in a specific offering and a potential intent to purchase. The RFI gives the vendor an opportunity to present new and changing information, and gives the recipient a better idea of current offerings.

An RFI can give a business information about several different vendors. By reviewing answers to an RFI, a network auditor or optimization specialist can enhance knowledge of changing offerings or gain knowledge of vendors the auditor does not normally work with. The RFI can be a very brief, informal document.

### Deciding on RFQ or RFP

An RFQ is a request for quotation. It does not ask the vendor for a formal proposal, it simply states requirements and asks for a price. An RFQ can save time for both the client and the vendor, but should only be used if the client has a long-term relationship with a specific vendor, and has a company direction specifying that only one vendor will be used.

When is an RFQ appropriate?

> If your company has a mandate that only one voice mail system will be used in all offices for proprietary voice mail networking.

> If your company has a mandate that only one brand of managed hub be used, to meet consistent network management requirements.

> If your company has decided to use a proprietary routing protocol, available only on one brand of networking equipment.

In short, you use a RFQ if you have already determined the hardware and software mix, and it is only a question of how much it costs to get another unit.

A company can also save time by issuing just one RFP a year for products of a like nature and purpose. This could include:

> One RFP a year for phone systems of a specific line and port size. The RFP would determine the vendor and system for the year. All additional purchases would be based on an RFQ.

> One RFP a year for routers using open routing protocols like OSPF. If the company used open protocols, there could be different brands of routers in the network, but they would use the same protocol. The RFP would determine the winning vendor for the year, additional purchases would be based on RFQs.

The time spent in creating a solid RFP and in analyzing it can be a great experience. But the money spent on having consultants write RFPs for like systems several times a year can be a waste.

Money spent on consultants for this purpose could easily pay for any trivial mistake made if you are dealing with ethical vendors. It would be less expensive to determine requirements once a year, and base all actions for the year on that one RFP.

The RFQ can be used in conjunction with an RFI. The RFI would give information about a general product set. The RFQ would give a price to provide the service. This can help a company determine if it should proceed with a formal RFP, or if the estimated costs of the project are too high.

While the RFQ can save time for both the vendor and the client, it leaves a lot of items open. The RFQ also does not give the vendor an opportunity to get to know client requirements well, and can lead to misunderstandings. An RFQ does not let the vendor offer the best product set to meet the client's requirements. The RFQ only provides a price for a set service.

A Request For Proposal allows the client to fully state his or her needs. The RFP should be distributed to multiple vendors, and vendor responses to the RFP should be weighed.

## Creating the RFQ

The Request For Quotation can be used to determine a price for a specific product. As vendors generally negotiate rates for non-tariffed services, the RFQ may not help a company obtain the most favorable price. The RFQ can be used to help reach a go/no go decision point.

The RFQ should contain at least a minimal fact set. This would include:
- Addresses of all the locations that will be connected to the network
- Area code & exchange for all offices
- Circuit speed requirements (for leased lines)
- Access speed, port speed and CIR requirements for Frame Relay
- Port mapping for frame relay
- Projected in service date

The RFQ should also request a price breakdown. This should be in one-time costs, and in monthly re-occurring charges.

## Creating the RFP

The Request For Proposal is a formal document requesting a specific service. It is used to narrow the field of vendors, and to determine the best product mix at the best price and performance for your company, and for your objectives.

The RFP should include all of your network requirements:

| | |
|---|---|
| Office locations | |
| Port speeds | |
| CIR | (if using Frame relay) |
| PVC mapping | (if using Frame Relay) |
| Responsibilities for Dmarc | (customer of vendor) |
| Responsibilities for CSU | (customer or vendor) |
| Reporting structures | |
| Availability | |
| Mean-time-to-repair | |

**Evaluating Proposals**

Evaluating proposals can be a time consuming task. Every vendor has different ways of saying the same thing. The task is:

    review all proposals
    determine differences in products
    determine differences in availability
    determine differences in network backbone
    determine required CIR :
        some vendors work well at OK CIR,
        other vendors dont work well even if the CIR equals port
           speed

It is critical to get all proposals on an equal footing, to make sure that you are evaluating the same service from each vendor. If services are different, you need to do a little "give and take" to make each proposal comparable.

After you are certain that all vendors are proposing the same network offering, the same port speeds, the same CIRs, and comparable services, you can start to see where proposed offerings differ.

Is the availability roughly equivalent? Did all vendors state that their offering has an uptime of 98.87-99.65% of the time? What is the difference in mean-time-to-repair, and how is it measured?

Do the vendors offer pro-active circuit, port and PVC assurance, or do you have to know when a device is down and call it in?

Do proposed CSUs (if you opted for vendor proposed CSUs) have multiple ports? If they do, you could transition dedicated voice services to the new carrier as your voice requirements grow.

If CSUs were proposed, do they offer testing facilities in both directions? Can the vendor tell if there is a problem with the circuit (testing towards the network), or if there is a problem with the router or cable (testing towards the equipment).

What type of reports are available, and what is the cost of the reports? Is a sample of the reports available, so you can determine the value of the reports?

## Selecting vendors

Vendor selection is more than simply a matter of which vendor has the lowest price. After reviewing all proposals and making sure that all vendors have bid on like services and like support, you can start to determine the winning vendor.

You should create a matrix, weighing which items are of most importance to your business. Each item would be given a weight, each vendors response would be reviewed to determine two things:

> how each vendor compares to other vendors
> how important the response is to your business

In reviewing vendors, it is helpful to create a simple spreadsheet. The spreadsheet would include each item from the proposal, and each vendor. You would enter the value of the response (i.e. Vendor 1 gets a 5, vendor 2 gets a 3, vendor 3 gets a 4, ...) and the weight of the response.

The spreadsheet can help you arrive at a numerical analyses of the top vendors. As some scores and weights may be subjective, you would then make a group decision on the winning vendor. This helps to keep personalities out of the equation.

# Part V - The Cutover

## - The Network Conversion

### Creating the Project Plan

After selecting a vendor, it is essential to create a project plan. The project plan must address each element of the network conversion.

As you build the project plan, the following items must be identified:

Who places circuit requests, customer or consultant

What is the lead time for circuits, by location

What is the signaling protocol (D4/AMI, ESF/B8ZS)

If using frame, what is the protocol - LMI or Annex-D

Are there any offices in areas that are experiencing excessive lead times?

Where is the Dmarc at each location

Who is responsible for extending the Dmarc to the computer room

Who supplies and configures the CSU/DSU?

When is the local loop due

When is the Dmarc scheduled to be extended to the computer room

When is the CSU scheduled to be installed and tested

Who accepts the circuit, and what constitutes acceptance

What will the CSU connect to?
> Router
> Bridge
> Gateway
> Control Unit, ...

What type of cable is required

Are spare cables available, or do they need to be ordered

If the circuit connects to a router, does the router have a spare serial port, or will it require a "hot cut"

If the router has a spare port, how long will the networks run parallel?

Who connects the cable from the CSU to the network equipment at each office?

When will the cable be connected

Is on-site support required, or can you talk the client through connecting the cable from the CSU to the network equipment.

When will disconnects be placed


**Placing Circuit Requests**
As the new network comes into being, it is important to keep up with circuit disconnects. There is normally a 30 day interval for disconnecting a circuit, so it is a judgement call on when to place disconnects.

The options depend on your budget and on how aggressive you are. Options include:
> Determine when new circuits should be live.
> Schedule conversion 3 to 5 days after circuits are in.
> Schedule disconnects to occur 15 days after circuits
> > are due.

OR

> Determine when circuits are due
> Connect equipment as soon as circuits are accepted
> Place disconnects ONLY after new circuits are in place and the new network has been tested and accepted.

The second alternative is obviously more expensive than the first alternative. The advantage of using the first alternative is increased savings, and having a definite timetable for converting the network. You can't let due dates slip if you have already placed disconnects!

While the second alternative is more expensive, it is also safer. If you are working with a critical network (and all of them are), you may opt for the second alternative to avoid any downtime.

This can be especially attractive if you run into provisioning problems. It is not uncommon for telco to be late delivering the local loop. This can be a matter of a few days, or several weeks (o longer).

There may also be problems with extending the Dmarc. The circuit can't come up without having the Dmarc extended. And finally, there may be problems with the CSU/DSU, or with the network equipment.

### Disruptive versus Non-Disruptive Conversion
A disruptive conversion requires the network to be down before a conversion can be made. If the networking equipment only has one serial port, and it is in use on the current network - it will need to be physically disconnected from the old CSU and then connected to the new CSU.

If you are working with routers, this may also require a change of the router configuration. While converting from one network carrier to another, you may have to have new TCP/IP and/or IPX/SPX networks assigned.

If the network equipment (router or bridge) has an available serial port, you can plan a non-disruptive conversion. By having an available serial ports and a spare V.35 cable, you will be able to connect the router to the the new network without disconnecting it from the old network.

A non-disruptive cutover can be more expensive:
    Spare CSU required
    Spare serial port required
    Spare cable required

But - a non-disruptive conversion can actually save money by not requiring network downtime. Every person that communicates across the wide area network would have a salary. By determining the number of people that will be idle, and the length of time they will be idle - you can determine the cost associated with bringing the network down to perform a conversion.

A rough example would be:
    24 people use the network for their job.
    They can not perform their job without the network
    Each person has a loaded salary of $90 an hour

    The loaded salary includes benefits, insurance, vacation, training, ...

    The disruptive conversion will take 30 minutes.
    .5 hours * 24 people * $90 an hour = $1,080 in lost wages.

This does not take into account any lost revenues due to the inability to process transactions for 30 minutes.

**Disruptive Conversion - Planning and Coordination**
Planning a disruptive conversion requires more coordination than planning a non-disruptive conversion. This is especially true if you need to change router configurations and you do not have the enable password to the router.

All conversions will require the following tasks:
    Install local loop
    Extend Dmarc
    Install CSU

The disruptive conversion will require the following tasks to be tightly coordinated:

Software Specialist:
> Dial into console port on router
> Verify networks and servers the router can see
> Load new configuration

Site Contact:
> Disconnect cable from old CSU
> Connect cable to new CSU

Software Specialist:
> Verify circuit is good
> (serial port should go active, or up/up)
>
> Verify frame mapping is good
> (make sure you can see correct end points)
>
> Make sure router can see same servers and networks it
>> previously had access to.

If all goes well, all tasks can be performed in 10 minutes. If there are any problems with the network (port not active on carrier's side, frame relay not mapped correctly), it may take longer.30 minutes should be reserved for the cutover.

This will give enough time to move the office to the new network, and to back out changes if things do not go as planned.

**Non-Disruptive - Planning and Coordination**
A non-disruptive cutover is a thing of beauty. It requires no downtime. It has a minimal effect on people in the office. It lets you run the old network and the new network in parallel.

A non-disruptive cutover involved many of the same things as a disruptive cutover. The local loop will need to be installed. The Dmarc may need to be extended. The new CSU will need to be installed and tested.

By shipping a spare cable to the office, the second serial port on the router can be connected to the new CSU while the primary serial port is connected to the old CSU.

By using a router with more than one serial port - all software changes can be rolled in ahead of time.

## The Fall-Back Plan - Disruptive Conversion
A fall-back plan for a disruptive cutover is a little more difficult than a non-disruptive cutover. The fall back includes:

> dial into the router
> disconnect the cable from the new CSU and circuit
> connect the V.35 cable to the old CSU
> reload the old configuration file
> check the serial port to make sure the port comes active
> verify frame relay mapping

The back-out plan should take 10 minutes to implement. As a person is required at the remote site to physically move the cable, and if you do not have access to make changes on the router - coordination will be required.

## The Fall-Back Plan - Non-Disruptive Conversion
The fall-back plan for a non-disruptive cutover is simple. All that is required is that the new serial port by shut down using a software command, and the old serial port be activated.

Both activating and inactivating ports is a simple software command. The procedure is as follows on a Cisco router:

> Telnet or dial into the router
> Enter the browse password
> Type Enable (or "en")
> Enter the enable password
> Type in conf t (for configure, terminal)
> Enter the serial port to be activated - i.e. S 0
> Type in "no shut" to activate the port
>
> Enter the serial port to be deactivate - i.e. S 1
> Type in "shutdown" to deactivate the port

After implementing the fall-back plan for either method, you could then call the circuit out to the carrier for testing, or report problems with the frame relay mapping.

## Placing Circuit Disconnects

Once the new circuits are in and tested, disconnects should be placed on the old circuits. All disconnect requests will need a minimum of the following information:

> Office address
> Site contact name and phone number
> Secondary contact name and number
> Circuit ID

A copy should be kept of all circuit disconnect requests. This will enable you to keep a record of when disconnect requests where placed.

After issuing disconnects, you should review network bills to make sure the circuits are removed from billing.

The final point of the network conversion is to verify billing. Make sure to have a copy of the initial proposal on file. If there were any questions over prices and responsibilities, make sure all replies are kept on file.

When you receive the bill for the new network, verify all costs.
> Make sure that port charges reflect proposed rates.
> Make sure that access charges are accurate
> Verify PVC charges

There have been occasions were actual prices on the first months bill do not reflect proposed prices at all. This can quickly add up to a substantial amount.

So:    Keep the proposal on file!
       Keep a hardcopy of all faxes and e-mails on file
       Make sure the bill reflects proposed services and prices!

# Part VI - Enterprise Network Considerations

An Enterprise network is simply a large, complex network serving a large corporation. The network will consist of varying communicating devices, using different protocols and supporting a mix of clients.

Enterprise network costs may be charged back to each department or organizational unit using the network.

The key points to consider in an enterprise network include:
> Verify what you are being billed for
> If you are being billed for a port - it should be active
> If you are billed for a controller, it should be active
> If you are billed for a circuit, it should be up
> If you are being billed for network devices, they should be
> > installed and in service.

The Enterprise Network billing group may be billing from the same information that the network control center and the helpdesk is using. If so - the information is almost gauranteed to be incorrect.


## Port Charges
If you are billed for port charges, you should be able to verify that the port is up, active and passing information. If not - the port charge should be removed.

As there is often a discrepancy between information sources and actual network equipment - all port charges should be verified at least once a year.


## Traffic Charges
As an enterprise network client, you may be billed for traffic going over the network. Traffic charges should be less than what it would cost to simply install your own leased lines or frame relay circuits.

The challenge is to determine what you are being charged, and to determine what it would cost to simply avoid the enterprise network.

## Management Charges

A central networking organization, or an Enterprise Network, may charge back management fees. This can consist of:

> HelpDesk support, 24x7
> Network Control Center, staffed 24x7
> Software specialists with advanced training
> Network design fees

If you have a fairly small network, it may be more cost effective to simply use the services of a central Enterprise Network team. Outsourcing (or insourcing) network services can allow you to concentrate on determining network directions, and growing towards true multi-media network communications.

By outsourcing the day-to-day network management chores to an Enterprise group, you will not need to build your own network control center. You also will not need the expense of staffing 24 hours a day, 7 days a week to meet online and batch requirements.

## Being a Good Enterprise Customer

What is a good Enterprise customer? The characteristics are:

> Notify HelpDesk  and NCC of all pending changes
> Allow time to plan and make network changes
> Create solid implementation plans
> Provide good back-out plans
> Provide disaster recovery plans
> Provide accurate documentation, and update as changes occur

# Appendix A

# Sample Network

Attached is our sample network. This is not meant as a network design exercise, but does show one way of maintaining information.

The network documentation also shows an evolution, from one carrier to a second, and then to a third. The document also shows how record keeping has changed. The first network has very little information. Records for carriers two and three are more detailed.

## Carrier A: The First Frame Provider

This document shows client sites and statuses for circuits. It can be noted that there is a large lag between disconnects as the network was moved to carrier two.

In our first iteration, we traveled to each office and physically moved them to a new carrier. This allowed tight quality control, as we were able to make all physical changes and verify connectivity.

This was also time consuming, and caused a major hit on travel and living expenses. We also only kept records for the last few offices to be migrated, not the entire 12 office network.

| Router | Status | Status Date | Circuit# | B/W | CIR | DLCI |
|--------|--------|-------------|----------|-----|-----|------|
| Client-A | Disconnected | Apr-96 | AREC35478.001 | 56 | 16 | 971 |
| HQ - #1 | Disconnected | Aug-96 | DZEC238911.200 | 256 | 64 | 317 |
| Client-B | Disconnected | Aug-96 | AREC35478.002 | 56 | 4 | 953 |
| Client-C | Disconnected | Jul-96 | AREC35478.003 | 56 | 16 | 958 |
| Client-D | Disconnected | Jul-96 | AREC35478.004 | 56 | 16 | 960 |
| Client-E | Disconnected | Jul-96 | AREC35478.005 | 56 | 16 | 969 |

A note about the included documentation: each spreadsheet also includes the monthly circuit costs (for 56K frame circuits), port costs and PVC costs. Access costs are kept for offices on Fractional T1.

Documentation is also kept on the expected length of time the circuit is expected to be kept in use. This lets us track monthly costs by office, and also lets us compare the bills to the proposal.

By tracking the length of time we expect the circuit to be up, we can determine what the office should pay per year for data communications. We can also have an idea of how much over budget we will be if a disconnect is pushed out.

## Carrier B: The Second Frame Provider
### Orders, Routers Ports, Status and Signaling

In reviewing the second carrier documentation, it is interesting to note that more information is kept. This allowed the client to start to integrate voice and data over the same access facilities.

Signaling formats also changed from D4/AMI to ESF/B8ZS. This did not effect network bill, but does allow port speeds in multiples of 64K, rather than multiples of 56K.

Also, the client changed from supplying CSUs, to having the vendor supply CSUs. While it may be more expensive to have the vendor supply the CSU, it also limits downtime and finger-pointing if a circuit goes down.

By having the telco extend the Dmarc to the computer room, rather than having a contractor extend the Dmarc, and by having the vendor supply the CSU - the client only has to call one help desk or control center when a circuit goes down.

The additional expenses are more than paid back by reducing the amount of time required to resolve a problem.

Why did the client move from carrier 1 to carrier 2? A higher speed network was required, and since all new facilities were required, it was a good point in time to compete the network.

# Carrier B: The Second Frame Provider

| Router | Port | Status | Order# | OE# | CSU | T1 Format | B/W | CIR |
|--------|------|--------|--------|-----|-----|-----------|-----|-----|
| HQ-#1 | S3/3 | up/up | | 266695 | Client | ESF/B8ZS | 1,024 | 0 |
| Client A | S0 | disc 8/96 | 880703 | 248739 | Client | D4/AMI | 112 | 0 |
| Client B | S0 | disc 8/96 | 884125 | 248864 | Client | D4/AMI | 112 | 0 |
| Client C | S0 | disc 8/96 | 884591 | 247696 | Client | D4/AMI | 112 | 0 |
| Client D | S0 | disc 8/96 | 884597 | 247704 | Client | D4/AMI | 112 | 0 |
| Client E | S0 | disc 8/96 | 884610 | 250312 | Client | D4/AMI | 112 | 0 |
| Client F | S0 | disc 8/96 | 884612 | 251492 | Vendor | D4/AMI | 112 | 0 |
| Client G | S0 | disc 8/96 | 884119 | 266057 | Client | ESF/B8ZS | 128 | 0 |
| Client H | S1 | disc 8/96 | 884850 | 265983 | Vendor | ESF/B8ZS | 128 | 0 |
| Client I | S1 | disc 8/96 | 884097 | 266690 | Vendor | ESF/B8ZS | 128 | 0 |
| Client J | S1 | disc 8/96 | 884849 | 266063 | Vendor | ESF/B8ZS | 128 | 0 |
| Client K | S1 | disc 8/96 | 884111 | 267011 | Vendor | ESF/B8ZS | 128 | 0 |
| Client L | S1 | disc 8/96 | 884114 | 265984 | Vendor | ESF/B8ZS | 128 | 0 |
| Client M | S1 | disc 8/96 | 884848 | 265981 | Vendor | ESF/B8ZS | 128 | 0 |
| Client N | S1 | up/up | (telco late) | 266051 | Vendor | ESF/B8ZS | 128 | 0 |
| Client O | S1 | disc 8/96 | | 266053 | Vendor | ESF/B8ZS | 128 | 0 |
| Client P | S1 | disc 8/96 | 884853 | 271288 | Vendor | ESF/B8ZS | 128 | 0 |
| Client Q | S1 | disc 8/96 | 884117 | 265978 | Vendor | ESF/B8ZS | 128 | 0 |
| Client R | S1 | disc 8/96 | 884129 | 265979 | Vendor | ESF/B8ZS | 128 | 0 |
| Client S | N/A | disc 8/96 | 884135 | 266058 | Vendor | ESF/B8ZS | 128 | 0 |
| Client T | S0 | disc 8/96 | 884130 | 281012 | Vendor | ESF/B8ZS | 128 | 0 |
| Client U | S0 | disc 8/96 | 884143 | 282430 | Vendor | ESF/B8ZS | 128 | 0 |
| HQ-#2 | N/A | disc 9/96 | | 292820 | Client | ESF/B8ZS | 1,544 | 0 |
| Client V | S0 | disc 8/96 | 884614 | 309857 | Vendor | ESF/B8ZS | 128 | 0 |
| | | | | | | | | |
| | | | | | | | | |
| Closed! | | | | | | | | |
| Client 1 | S0 | disc 4/96 | 837142 | 247685 | Client | D4/AMI | 112 | 0 |
| Client 2 | S0 | disc 4/96 | 837151 | 247812 | Client | D4/AMI | 112 | 0 |
| Client 3 | S0 | disc 3/96 | 808712 | 251692 | Vendor | D4/AMI | 112 | 0 |
| Client 4 | S0 | disc 4/96 | 852397 | 247689 | Client | D4/AMI | 112 | 0 |
| Client 5 | S1 | disc 5/96 | 846437 | 270566 | Vendor | ESF/B8ZS | 128 | 0 |
| Client 6 | S0 | disc 8/96 | 850588 | 315915 | Vendor | ESF/B8ZS | 128 | 0 |

## Carrier B: The Second Frame Provider - Circuit IDs

| Router | PL# | Local Loop | DLCI | Net Addr |
|---|---|---|---|---|
| HQ-#1 | 191869 | HCGS621331 | 421 | 44179135 |
| Client A | 183329 | 11HCGS204602 | 422 | 44143048 |
| Client B | 183452 | 12HCGS204602 | 423 | 44151007 |
| Client C | 183029 | HCGS407221 | 424 | 44144085 |
| Client D | 183041 | 49HCGS030000488 | 426 | 44120152 |
| Client E | 184270 | HCGS069018 | 430 | 44158120 |
| Client F | 184845 | 41HCGS604523 | 439 | 44190098 |
| Client G | 191617 | HCGS621331 | 449 | 44179134 |
| Client H | 191522 | 38HCGS617075 | 450 | 44201015 |
| Client I | 191873 | 11HCGS205911 | 451 | 44179136 |
| Client J | 191573 | 38HCGS770849 | 452 | 44153036 |
| Client K | 192063 | CLT300893 | 453 | 44202007 |
| Client L | 191594 | 14HCGS645342 | 454 | 44142278 |
| Client M | 191517 | TDL00490601 | 455 | 44154014 |
| Client N | 191615 | 40HCGS030004972 | 456 | 44172059 |
| Client O | 191611 | 82HCGS030002706 | 457 | 44137156 |
| Client P | 193864 | 70HCGS030005181 | 458 | 44206062 |
| Client Q | 191584 | HCGS729390 | 459 | 44140212 |
| Client R | 191515 | 15HCGS47173 | 460 | 44174184 |
| Client S | 191581 | 75HCGS780239 | 461 | 44133150 |
| Client T | 198115 | 01ATTZ005005026 | 466 | 44184151 |
| Client U | 198788 | 86HCGS401621 | 468 | 44150157 |
| HQ-#2 | 203004 | 38HCGS619060 | 485 | 44252005 |
| Client V | 209843 | 38HCGF62080 | 492 | 44273021 |
| Client 1 | 183012 | 69HCGS103793 | 428 | 44131184 |
| Client 2 | 183063 | HCGS068827 | 429 | 44119246 |
| Client 3 | 184914 | 84HCGS030000212 | 438 | 44134244 |
| Client 4 | 183035 | 744HCGS904997 | 425 | 44190079 |
| Client 5 | 193654 | 73HCGS030005169 | 462 | 44137160 |
| Client 6 | 212348 | 95HCGS020940 | 491 | 44296013 |

Why did the client move from carrier 2? There were no service issues. The second carrier had good performance, good project management, and a solid network offering.

As frame relay is becoming more of a commodity, it is being priced more like a commodity. Frame relay can be provided by several RBOCs and IXCs. This increased competition has reduced costs for frame relay networks significantly.

As carriers move quickly to compete in local and long distance, and to compete in data communications - the savings potential will only become even greater.

The last point about the conversion to carrier three is the speed in which the conversion took place.

All orders were placed with 30 day lead times.
All software requests were made when circuit orders placed.
All routers were reviewed for spare serial ports.
Additional cables were ordered for each office.
The telco extended Dmarcs.
The carrier installed CSUs.
The carrier connected router cables to CSUs in several locations
Specific instructions were provided to each office, where the vendor did not connect cables, the office staff did.
NO trips were required!
The Travel and Living budget was untouched!
21 out of 22 offices had a non-disruptive cutover
1 office had to be stepped through the process, as they only had 1 serial port on the router
All offices were up in parallel, offering a simple conversion, and a painless back-out plan

## Carrier C: The Third (and final?) Frame Provider
### - Order #s, Speed, Status, T1 IDs

| Router | Port | Status | Order# | Service# | CSU | T1 ID | Chnl | CCT Speed | CIR |
|--------|------|--------|--------|----------|-----|-------|------|-----------|-----|
| HQ - #1 | S3/6 | up/up | ID912052 | P0388175 | Vendor | 188785 | 1-24 | 1.544 Mbps | 0K |
| HQ - #2 | N/A | up/up | ID872341 | P0385387 | Client | 180440 | 1-24 | 1.544 Mbps | 0K |
| Client A | S1 | up/up | ID867982 | P0385040 | Vendor | 180265 | 01,02 | 128 Kb | 0K |
| Client B | S1 | up/up | ID870183 | P0385228 | Vendor | 180372 | 01,02 | 128 Kb | 0K |
| Client C | S1 | up/up | ID882823 | P0386138 | Vendor | 180810 | 01,02 | 128 Kb | 0K |
| Client D | S1 | up/up | ID882932 | P0386145 | Vendor | 180813 | 01,02 | 128 Kb | 0K |
| Client E | S1 | up/up | ID870646 | P0385281 | Vendor | 180397 | 01,02 | 128 Kb | 0K |
| Client F | S1 | up/up | ID899340 | P0387360 | Vendor | 188251 | 01,02 | 128 Kb | 0K |
| Client G | S0 | up/up | ID873323 | P0385421 | Vendor | 180451 | 01,02 | 128 Kb | 0K |
| Client H | S0 | up/up | ID873496 | P0385443 | Vendor | 180457 | 01,02 | 128 Kb | 0K |
| Client I | S0 | up/up | ID883999 | P0386248 | Vendor | 180907 | 01,02 | 128 Kb | 0K |
| Client J | S0 | up/up | ID886629 | P0386420 | Vendor | 181053 | 01,02 | 128 Kb | 0K |
| Client K | S0 | up/up | ID901197 | P0387462 | Vendor | 188341 | 01,02 | 128 Kb | 0K |
| Client L | S0 | telcol | ID882298 | P0386093 | Vendor | 180775 | 01,02 | 128 Kb | 0K |
| Client M | S1 | up/up | ID887059 | P0386463 | Vendor | 181078 | 01,02 | 128 Kb | 0K |
| Client N | S0 | up/up | ID890503 | P0386683 | Vendor | 187841 | 01,02 | 128 Kb | 0K |
| Client O | S0 | up/up | ID890524 | P0386686 | Vendor | 187843 | 01,02 | 128 Kb | 0K |
| Client P | S0 | up/up | ID899764 | P0387389 | Vendor | 188267 | 01,02 | 128 Kb | 0K |
| Client Q | S0 | up/up | ID870754 | P0385296 | Vendor | 180401 | 01,02 | 128 Kb | 0K |
| Client R | S1 | up/up | ID866794 | P0386442 | Vendor | 181062 | 01,02 | 128 Kb | 0K |
| Client S | S1 | up/up | ID908210 | P0387909 | Vendor | 188648 | 01,02 | 128 Kb | 0K |
| Client T | S1 | up/up | ID901214 | P0387464 | Vendor | 188350 | 01,02 | 128 Kb | 0K |
| Client U | S1 | up/up | ID871179 | P0385309 | Vendor | 180404 | 01,02 | 128 Kb | 0K |
| Client V | S0 | up/up | ID966755 | P0392796 | Vendor | 191181 | 01,02 | 128 Kb | 0K |

## Carrier C: The Third (and final?) Frame Provider - Circuit IDs

| Router | MCI Ckt ID | Telco ID | DLCI |
|---|---|---|---|
| HQ - #1 | ZA U46538 0001 | /HCGS/628449//SN/ | 40 |
| HQ - #2 | ZA U31423 0001 | DS3#3, Ch2 | 30 |
| Client A | ZA U29901 0001 | 11/HCGS/210421//PA | 100 |
| Client B | ZA U30820 0001 | 13-D49-FSA-0001 | 101 |
| Client C | ZA U35241 0001 | HCGS/410728//CB/ | 102 |
| Client D | ZA U35281 0001 | 89/HCGS/005008/251/PT | 103 |
| Client E | ZA U30971 0001 | HCGS/078201//NJ/ | 104 |
| Client F | ZA U40852 0001 | 41/HCGS/606325//SW/ | 105 |
| Client G | ZA U31728 0001 | 38/HCGS/622392//SB/ | 106 |
| Client H | ZA U31780 0001 | 11/HCGS/210420//PA/ | 107 |
| Client I | ZA U35731 0001 | 16/HCGS/00412//TW | 108 |
| Client J | ZA U36690 0001 | 13-D49-EJJ-0001 | 109 |
| Client K | ZA U41579 0001 | 24/HCGS/589410//MS/ | 110 |
| Client L | ZA U35068 0001 | 40/HCGS/005009/778/PT | 111 |
| Client M | ZA U36815 0001 | 70/HCGS/005009/730/PT | 112 |
| Client N | ZA U37811 0001 | 74/HCGS/005159/314/PT | 113 |
| Client O | ZA U37816 0001 | HCGS/736736//LB/ | 114 |
| Client P | ZA U41037 0001 | 15/HCGS/49045//NW/ | 115 |
| Client Q | ZA U31005 0001 | 50/HCGS/202709//UTNJ | 116 |
| Client R | ZA U36755 0001 | 32/HCGS/176435//NY/ | 117 |
| Client S | ZA U44400 0001 | 86/HCGS/402098//GTEW/ | 118 |
| Client T | ZA U41581 0001 | 38/HCGS/622388/SB/ | 119 |
| Client U | ZA U31122 0001 | HCGS/628138//SN/ | 120 |
| Client V | ZA U72276 0001 | 38/HCGS/772073 | 121 |

## Router Ports:

The following chart shows router serial ports (and IP addresses), and what carrier the port is connected to. Port status is also tracked - to make sure disconnects have taken place, and the new installations are progressing.

New installations progressed rapidly! Disconnects took place quickly!

| Router | RtrStatus | Serial_0 | Carrier | Status | Serial_1 | Carrier | Status |
|--------|-----------|----------|---------|--------|----------|---------|--------|
| HQ-#1 | production | 3.119.124.1 | #2 | up | 3.155.88.1 | #3 | up |
| Client A | production | 3.119.124.8 | #2 | down | 3.155.88.3 | #3 | up |
| Client B | production | 3.119.124.9 | #2 | down | 3.155.88.4 | #3 | up |
| Client C | production | 3.119.124.2 | #2 | down | 3.155.88.5 | #3 | up |
| Client D | production | 3.119.124.6 | #2 | down | 3.155.88.6 | #3 | up |
| Client E | production | 3.119.124.10 | #2 | down | 3.155.88.7 | #3 | up |
| Client F | production | 3.119.124.11 | #2 | down | 3.155.88.8 | #3 | up |
| Client G | production | 3.119.124.25 | #2 | down | 3.155.88.23 | #3 | up |
| Client H | production | 3.155.88.9 | #3 | up | 3.119.124.14 | #2 | down |
| Client I | production | 3.155.88.10 | #3 | up | 3.119.124.15 | #2 | down |
| Client J | production | 3.155.88.24 | #3 | up | 3.119.124.13 | #2 | down |
| Client K | production | 3.155.88.11 | #3 | up | 3.119.124.16 | #2 | down |
| Client L | production | 3.155.88.12 | #3 | up | 3.119.124.17 | #2 | down |
| Client M | production | 3.155.88.13 | #3 | up | 3.119.124.18 | #2 | down |
| Client N | production | 3.155.88.14 | #3 | up | 3.119.124.19 | #2 | down |
| Client O | production | 3.155.88.16 | #3 | up | 3.119.124.20 | #2 | up |
| Client P | production | 3.83.156.9 | #1 | down | 3.155.88.15 | #3 | up |
| Client Q | production | 3.155.88.17 | #3 | up | 3.119.124.22 | #2 | down |
| Client R | production | 3.155.88.15 | #3 | up | 3.119.124.23 | #2 | down |
| Client S | production | 3.155.88.19 | #3 | up | (no s1 port) | | |
| Client T | production | 3.119.124.27 | #2 | down | 3.155.88.20 | #3 | up |
| Client U | production | 3.119.124.26 | #2 | down | 3.155.88.21 | #3 | up |
| Client V | production | 3.119.124.29 | #2 | down | 3.155.88.22 | #3 | up |
| HQ - #2 | production | 3.119.124.30 | #2 | down | 3.155.88.2 | #3 | up |

## Routers: Ethernet ports, Memory and Software Levels

By keeping track of router memory and software levels, we can determine when a router is back-level. Several are in the sample network, although they all do the job required of them at the software level they are running.

By tracking memory, we can determine if memory upgrades are required to support current software levels. The next step is to price out memory, have the memory installed, then upgrade software to current versions.

| Router | Ethernet_0 | Ethernet_1 | Model | Memory | Software |
|--------|-----------|-----------|-------|--------|----------|
| HQ-#1 | (hq router) | | 7500 | | 10.3(8) |
| Client A | 3.119.152.1 | (no e1 port) | 2500 | 16380K/2048K | 10.0(8) |
| Client B | 3.119.156.1 | (no e 1 port) | 2500 | 16380K/2048K | 10.0(8) |
| Client C | 3.119.128.1 | (no e 1 port) | 2500 | 1024K/1024K | 10.2(5) |
| Client D | 3.119.144.1 | (no e1 port) | 2500 | 16380K/2048K | 10.0(8) |
| Client E | 3.119.160.1 | (no e 1 port) | 2500 | 16380K/2048K | 10.0(8) |
| Client F | 3.119.164.1 | 3.123.152.1 | 4500 | 32768K/4096K | 10.0(2) |
| Client G | 3.97.228.1 | (no e 1 port) | 2500 | 16380K/2048K | 10.0(8) |
| Client H | 3.99.244.1 | 3.99.248.1 | 4000 | 4096K/4096K | 9.14(3) |
| Client I | 3.99.228.1 | 3.99.232.1 | 4000 | 4096K/4096K | 9.14(3) |
| Client J | 3.88.16.1 | 3.83.252.1 | 4000 | 4096K/4096K | 9.14(8) |
| Client K | 3.99.236.1 | 3.99.240.1 | 4000 | 4096K/4096K | 9.14(8) |
| Client L | 3.99.72.1 | 3.99.76.1 | 4000 | 4096K/4096K | 9.14(3) |
| Client M | 3.99.12.1 | 3.99.16.1 | 4000 | 4096K/4096K | 9.14(3) |
| Client N | 3.83.160.1 | 3.113.36.1 | 4000 | 8192K/4096K | 9.14(8) |
| Client O | 3.99.4.1 | 3.99.8.1 | 4000 | 4096K/4096K | 9.14(3) |
| Client P | 3.99.80.1 | 3.99.84.1 | 4000 | 4096K/4096K | 9.14(3) |
| Client Q | 3.99.164.1 | 3.99.168.1 | 4000 | 4096K/4096K | 9.14(7) |
| Client R | 3.88.20.1 | 3.88.164.1 | 4000 | 4096K/4096K | 9.14(5) |
| Client S | 3.97.200.1 | (no e 1 port) | 3000 | 16384K/512K | 9.1(7) |
| Client T | 3.79.236.1 | (no e 1 port) | 2500 | 16380K/2048K | 10.0(6) |
| Client U | 3.131.236.1 | (no e 1 port) | 2500 | 8188K/2048K | 10.2(5) |
| Client V | 3.150.96.1 | (no e1 port) | 2500 | 16380K/2048K | 10.0(8) |
| HQ - #2 | | | | | |

The dial-in number for each router is also kept, as well as the site contact and phone number/pager number. Router costs, maintenance costs, and purchase date should also be maintained.

## ISDN Routers for Disaster Recovery:

Every office that is on the frame relay network also has a disaster recovery platform. Each office has a small ISDN router. In the event that the primary router takes a nose-dive, the backup router is on site and powered up.

If the circuit goes down, the ISDN router dials into a PRI on a router at the Headquarters office if any devices on the LAN have data to transmit. This is dial-on-demand routing. Additionally, the HQ router will dial out to the backup router using ISDN if the primary route is down.

The ISDN routers required some "fooling". We built floating static TCP/IP and IPX/SPX routes. We also built network filters and SAP filters for each office, and faked the hop count to make it look like the dial-platform was a long distance away. That way, dial-up only occurs when frame is down.

## ISDN Routers - Serial and Ethernet IP and IPX Addresses:

| Router | Serial-IP | Serial-IPX | Eth-IP | Eth-IPX |
|---|---|---|---|---|
| bri-client A | 3.150.12.11 | 3960c00 | 3.119.152.2 | 3779800 |
| bri-client B | 3.150.12.12 | 3960c00 | 3.119.156.2 | 3779C00 |
| bri-client C | 3.150.12.13 | 3960c00 | 3.119.128.2 | 3778000 |
| bri-client D | 3.150.12.14 | 3960c00 | 3.119.144.2 | 3779000 |
| bri-client E | 3.150.12.15 | 3960c00 | 3.119.160.2 | 377A000 |
| bri-client F | 3.150.12.16 | 3960c00 | 3.119.164.2 | 3777C00 |
| bri-client G | 3.150.12.17 | 3960c00 | 3.97.228.2 | 361E400 |
| bri-client H | 3.150.12.18 | 3960c00 | 3.99.244.2 | 363F400 |
| bri-client I | 3.150.12.19 | 3960c00 | 3.99.228.2 | 363E400 |
| bri-client J | 3.150.12.20 | 3960c00 | 3.88.16.2 | 3581000 |
| bri-client K | 3.150.12.21 | 3960c00 | 3.99.236.2 | 363EC00 |
| bri-client L | 3.150.12.22 | 3960c00 | 3.99.72.2 | 3634800 |
| bri-client M | 3.150.12.23 | 3960c00 | 3.99.12.2 | 3630C00 |
| bri-client N | 3.150.12.24 | 3960c00 | 3.83.160.2 | 3831600 |
| bri-client O | 3.150.12.25 | 3960c00 | 3.99.4.2 | 3630400 |
| bri-client P | 3.150.12.26 | 3960c00 | 3.99.80.2 | 3635000 |
| bri-client Q | 3.150.12.27 | 3960c00 | 3.99.164.2 | 363A400 |
| bri-client R | 3.150.12.28 | 3960c00 | 3.88.20.2 | 358140 |
| bri-client S | 3.150.12.29 | 3960c00 | 3.97.200.2 | 361C800 |
| bri-client T | 3.150.12.30 | 3960c00 | 3.79.236.2 | 34FEC00 |
| bri-client U | 3.150.12.31 | 3960c00 | 3.131.236.2 | 383EC00 |
| bri-client V | 3.150.12.32 | 3960c00 | 3.150.96.2 | 3966000 |

## ISDN Routers:
## IPX MAC addresses, serial numbers, model numbers

| Router | IPX Addr | Serial | Model | IOS | Feature |
|---|---|---|---|---|---|
| bri-client A | 0000.0c32.c390 | 02044577 | 1004 | 10.3(6) | IP/IPX/AT |
| bri-client B | 0000.0c32.a628 | 02056352 | 1004 | 10.3(6) | IP/IPX/AT |
| bri-client C | 0000.0c32.c32f | 02073754 | 1004 | 10.3(6) | IP/IPX/AT |
| bri-client D | | | 1004 | 10.3(6) | IP/IPX/AT |
| bri-client E | 0000.0c32.c344 | 02075647 | 1004 | 10.3(6) | IP/IPX/AT |
| bri-client F | 0000.0c32.c399 | 02044593 | 1004 | 10.3(6) | IP/IPX/AT |
| bri-client G | 0000.0c32.a813 | 02062449 | 1004 | 10.3(6) | IP/IPX/AT |
| bri-client H | 0000.0c32.a608 | 02059891 | 1004 | 10.3(6) | IP/IPX/AT |
| bri-client I | 0000.0c32.eb44 | 02045345 | 1004 | 10.3(6) | IP/IPX/AT |
| bri-client J | 0000.0c32.a826 | 02062452 | 1004 | 10.3(6) | IP/IPX/AT |
| bri-client K | 0000.0c32.c394 | 02044553 | 1004 | 10.3(6) | IP/IPX/AT |
| bri-client L | 0000.0c32.ebae | 02043963 | 1004 | 10.3(6) | IP/IPX/AT |
| bri-client M | 0000.0c32.a641 | 02056298 | 1004 | 10.3(6) | IP/IPX/AT |
| bri-client N | 0000.0c32.ebae | 02043963 | 1004 | 10.3(6) | IP/IPX/AT |
| bri-client O | 0000.0c32.eb46 | 02045542 | 1004 | 10.3(6) | IP/IPX/AT |
| bri-client P | 0000.0c32.eb45 | 02043926 | 1004 | 10.3(6) | IP/IPX/AT |
| bri-client Q | 0000.0c32.a5e0 | 02055821 | 1004 | 10.3(6) | IP/IPX/AT |
| bri-client R | 0000.0c32.a5e5 | 02055794 | 1004 | 10.3(6) | IP/IPX/AT |
| bri-client S | 0000.0c32.a63b | 02056295 | 1004 | 10.3(6) | IP/IPX/AT |
| bri-client T | 0000.0c32.a5e2 | 02055867 | 1004 | 10.3(6) | IP/IPX/AT |
| bri-client U | 0000.0c32.a68a | 02059950 | 1004 | 10.3(6) | IP/IPX/AT |
| bri-client V | 0000.0c32.a823 | 02062446 | 1004 | 10.3(6) | IP/IPX/AT |

Router passwords for PPP authentication CHAP are also kept, as well as a spreadsheet with all ISDN phone numbers, circuit IDs, CO switches, ...

# Appendix B

# Network Auditing Toolkit

While a network auditor may be able to reduce costs by reviewing network documentation and network bills, the auditor can further reduce costs or provide a consulting service by verifying network connectivity and trending network utilization over a set period of time.

This network trending will give the client a baseline of network utilization, and allow the client to plan for increased network speeds as it is required. Whle the client may have an SNMP management platform that checks to see if network devices are active, the client may not have a network trending tool.

The tool kit will allow the auditor to verify what physical devices are on the network. By focusing the view on the wide area network, the auditor can provide a service that is important to the client.

The network audit will show what circuits are in use, and what circuits can be disconnected. The tool kit will allow the auditor to determine what portion of the circuit speed is being used. A constantly low circuit utilization level will provide information that the circuit speed should be reduced. This will result in a cost savings as vendors bill on circuit speed.

The tool kit will also allow the auditor to determine what PVCs are in use, and what PVCs are no longer required. By removing under-utilized PVCs, the monthly bill will be further reduced. Intermittent traffic could be routed from one office to another via the serving router.

By trending CIR utilization, the auditor will be able to determine if the client requires a specific CIR. Frame relay vendors bill on CIR speed. Some vendors operate well at 0K CIR, and burst all traffic. Some vendors require a set CIR, but offer several options in CIR speed. Some vendors don't burst well at all, and a relatively high CIR may be required. By trending utilization, the auditor will be able to determine what the CIR should be.

ISDN is another area that should be watched. By trending ISDN usage, the auditor will be able to determine how many ISDN BRI circuits are required at a central site to serve dial-up routers and bridges. Trending utilization will give the auditor the ability to determine if more ISDN BRI circuits or ISDN PRI circuits are required. Trending utilization will also let the auditor know if ISDN circuits should be disconnected.

**The tool kit:**
The tool kit should be relatively inexpensive, as hardware and software sometimes walks. Before placing equipment on a customer's premises, you would need to determine responsibility for the equipment.

Network Management applications are continuously evolving. Applications that are strong at the time of this writing, may be replaced in the near future. Applications that are written for the Unix platform may be ported to Windows NT in the near future.

As with Traffic Engineering and Telemanagement applications for voice systems, there is a wide variety of features in various network management applications.

The main items the network auditor will require include:

> **Wide Area Network (WAN) Support**
> These measurements should be both real-time and historical, allowing trend analyses.
>
> Leased line utilization
> Frame Relay utilization
> > Port Speed utilization
> > CIR utilization
> > PVC utilization
> ISDN PRI
> > Number of calls in time period
> > Average length of call
> > Busy hour stats
> > Maximum simultaneous calls
> ISDN BRI
> > Number of calls in time period
> > Average length of call
> > Busy hour stats
>
> Routers
> The tool should be able to measure Cisco and Wellfleet routers. These are the most popular. It would be good to measure other

routers as well. This is analogous to having a tool set that measures AT&T/Lucent switches, Northern Telecomm switches, and Siemens/Rolm switches. It would be good to be able to measure other switches, but the cost of doing so may not pay for the tools required to do the job.

**Protocols:**
The tool set should support TCP/IP and IPX/SPX protocols at a minimum. It would be good to be able to measure AppleTalk as well.

One option for trending wide area network utilization is Stony Brook Software's WAN Manager product. This is an application that runs on a PC running Windows 95. The PC could be a relatively inexpensive Laptop or portable, that could be easily transported from one client to another. Stony Brook's WAN specific product lists for $5K.

There are less expensive alternatives as well. If you do not require a Windows platform, there are products available at little or no cost that run on the Unix platform. A good place to look for products is the comp.dcom.net-management newsgroup. There is a FAQ available. The FAQ points to a Web server that lists current products, and offers downloads of products based on perl scripts.

# Appendix C: Sample Proposal

A proposal consists of the following items.

1. Cover letter
2. Title page
3. Table of contents
4. Executive summary
5. Body of Proposal
     5A. Technical detail
     5B. Deliverables
     5C. Cost justification and return on investment (ROI)
     5D. Schedules
     5E. Documentation
     5F. Resumes
6. Glossary
7. Appendix

The details of your proposal will vary, depending on the complexity and competitiveness of the job. Following is a sample proposal that you may modify to meet your needs. A proposal can be used as a legal document, so think clearly before you submit a proposal. Do not exaggerate

# Elizabeth-Bruce Communications Consultants
## 30 Angela Court, Beacon NY 12508

[date]

Prospect Name
Prospect Address
Prospect City, State, ZIP

Dear Mr. Prospect:

We are pleased to submit the enclosed proposal in response to your recent inquiries concerning telecommunications auditing.

*Elizabeth-Bruce Communications Consultants* provides a unique service, designed to meet the voice and data communications auditing, network optimization, and documentation needs of your company.

We propose to review all data communications invoices to identify all circuits currently being billed to you. We will verify that all circuits being billed are in place and in use. We will notify you of all circuits not currently in use, and will issue disconnect requests upon your approval.

We will monitor your routed Wide Area Network utilization over a period of 45 days to determine appropriate circuit speeds. We will make recommendations to increase or decrease circuit speeds as required. If you choose, we will work directly with your network staff, your remote offices, and your network provider to project manage recommended network changes.

*Elizabeth-Bruce Communications* will monitor dial-in access to your E-Mail server, Asynchronous Terminal Server and ISDN router to determine the appropriate number of modems and trunks required to serve your dial-in clients. We will make a recommendation based on the average number of busy signals out of every 100 attempts you wish to give your clients. We will determine calling patterns, and recommend low cost alternatives for dial access.

---

*Elizabeth-Bruce Communications* will provide complete documentation, consisting of the following items:

> Circuits in use
> Cost per circuit
> Carrier
> Carrier Escalation Levels
>
> Router Network Equipment in place
> Connecting circuits
> TCP/IP and/or IPX/SPX addresses
> Router software level
> Router memory
> Dial-in number to console port
>
> IBM SNA/SDLC Equipment
> Connecting circuits
> Cost per circuit
>
> Primary and Secondary Remote Site Contacts
> Phone numbers
> Fax numbers
> Pager numbers
> Cellular numbers
> E-Mail addresses

Finally, *Elizabeth-Bruce Communications* will update your information management systems, allowing your network staff to quickly respond to network outages with accurate information, contacts and escalation levels.

Sincerely,

Bruce W. Griffis
President, Elizabeth-Bruce Communications Consultants

# Reduce Network Costs And Improve

## *Network Response Time*
## *Network Performance*
## *and*
## *Network Availability*

A proposal from
Elizabeth-Bruce Communications Consulting
to
Prospect Company

[date]

# Executive Summary

*The bottom line is:*
> *Elizabeth-Bruce can reduce your network costs!*
> *Elizabeth-Bruce can improve your network response times!*
> *Elizabeth-Bruce can improve your network performance!*
> *Elizabeth-Bruce can improve your network availability!*

The world of telecommunications has evolved rapidly, growing from low speed analog circuits in hierarchical networks to high speed digital circuits supporting client-server applications.

As networks evolve, it is crucial to integrate legacy networks with new multi-protocol networks, rather than building parallel networks. It is crucial to disconnect circuits that are no longer in use, and it is critical to monitor network utilization to determine the best place to invest your networking dollars.

Networks expand (at increased costs) to meet new application requirements and shifts in personnel. Networks often do not contract as requirements decrease. Contraction is in the form of decreased circuit speeds, decreased port speeds, and decreased CIR. Contraction is also in the form of disconnected PVCs when offices no longer exchange information in a peer to peer environment.

By focusing on changing requirements, *Elizabeth-Bruce* can help you reduce network expenses where higher speeds are no longer necessary. Elizabeth-Bruce can also point out where investments in increased network capacity are warranted.

How can *Elizabeth-Bruce* reduce your network costs?

1.  By verifying circuit inventories
2.  By disconnecting unused circuits
3.  By using network measurement tools to determine appropriate circuit speeds, port speeds, CIRs and PVCs

---

How can *Elizabeth-Bruce* improve your network response times?

1. By increasing circuit speeds where appropriate
2. By increasing port speeds where appropriate

How can *Elizabeth-Bruce* improve network performance?

1. By measuring network throughput
2. By increasing CIR to reduce dropped bits
3. By adding PVCs as application servers and application clients move

How can *Elizabeth-Bruce* improve network availability?

1. By documenting the physical network
2. By documenting logical network connectivity
3. By providing the Help Desk with valid office contacts, phone numbers and pagers
4. By providing the Help Desk with proper escalation levels with network providers
5. By providing remote offices with proper procedures for reporting network outages

*Elizabeth-Bruce* can assist your help desk and network control center while saving you money. In many cases your annualized savings will pay for our services, while enabling you to build expertise in-house for future network optimization projects.

# Technical Overview

*Elizabeth-Bruce Communications Consultants* takes a holistic approach to network auditing. We concentrate on the components that comprise the wide area network.

Elizabeth-Bruce will perform a **circuit inventory**. We will review all circuit bills and Customer Service Records (CSRs). We will compare circuits noted in your Information Management system to circuits that appear on your bills. We will document all anomalies, as a first step in auditing the wide area network.

We will perform a site survey at larger locations and corporate computer centers to verify all circuits currently in place. We will trace circuit connectivity to actual network hardware, to ensure that all circuits appearing on circuit bills are actually connected to network equipment.

We will update your Information Management system with physical circuit connectivity, including circuit name, location, and connecting hardware. We will place disconnect orders for all circuits not in use, upon verification from your network staff.

We will verify the following circuit related items:
>
> Circuit ID
> Port Speed
> Committed Information Rate (for Frame Relay)
> PVC Mapping
>
> Access cost - related to Circuit ID
> Port cost
> CIR cost
> PVC costs
>
> Carrier
> Carrier's problem reporting number
> Carrier's escalation procedures

*Elizabeth-Bruce Communications Consultants* will perform a **Network Equipment Inventory**. We will verify all networking equipment that touches the wide area network. This will include routers, bridges, and legacy equipment.

The Network Equipment Inventory consists of both a physical inventory and a logical inventory. We will perform both a physical and logical inventory of network equipment at larger offices and computer centers. We will perform a logical inventory at smaller offices.

> **Physical Inventory**: we will address the following issues during the physical inventory:
>
>> Serial port connectivity
>>> (local router, circuit ID, remote router)
>>
>> Asynchronous modem number (for console)
>> Serial number
>
> **Logical Inventory**: we will address the following issues during a logical inventory:
>
>> Equipment manufacturer
>> Equipment model
>>
>> Serial port status (active, inactive)
>> Serial port connectivity
>>> Local router -> CircuitID-> Remote router
>>
>> WAN connectivity:
>>> Frame Relay
>>> Leased Line
>>> ISDN
>>
>> Memory
>> Software level
>>
>> Maintenance contract
>> Vendor's support number

After performing a network equipment inventory, we will update your Information Management systems. This will include current networking equipment and serial numbers, connecting circuits, and modem numbers for out-of-band management.

---

As personnel changes over time, and Information Management systems do not always reflect those changes, we will **verify all site contacts**. We will review current site contacts based on Information Management records. We will then call all remote offices to verify contacts.

This portion of the audit will include:
    Verify primary site contact
    Verify secondary site contact

    Verify phone number, fax number, e-mail address and pager number for both contacts.

    Document on-call responsibilities for network outages after normal business hours.

    Verify street address and computer room location.

Upon completion, we will update your Information Management system, to include validated contacts at all locations. We strongly feel that site contacts should be validated at least twice a year, if not every quarter.

After reviewing circuit inventories, Wide Area Network inventories, and verifying site contacts, *Elizabeth-Bruce Communications Consultants* will create a **logical network map**.

    This map will consist of:
    WAN equipment
        Routers, Bridges, Legacy Equipment
        Manufacturer, model
    Circuits
        Circuit Speeds, Port Speeds, CIR, Circuit IDs
    Physical locations
    Contacts

If you have an Intranet Web Server, *Elizabeth-Bruce Communications Consultants* will create a logical network map in appropriate formats (HTML, GIF and JPEG). This map can be posted on your Intranet server, as a means of providing graphical network documentation for your help desk and network staff.

Graphical documentation can simplify the problem determination and problem resolution time, by showing exactly what Wide Area Networking components are connected, and documenting the circuits that connect them.

We will post vendor, carrier, and site contacts on your Intranet Web Server. This will reduce the time required to find the correct phone numbers and begin problem resolution.

The final aspect of network auditing and documentation is monitoring **network utilization**. *Elizabeth-Bruce Communications Consultants* uses two forms of network performance monitoring, based on your requirements. Our primary goal is to monitor the Wide Area Network to determine bandwidth requirements.

As Wide Area Network circuits comprise a major portion of telecommunications budgets, it is important to understand the billing aspects of circuits. It is also important to determine how you are using what you are billed for.

**Leased Lines:**
Leased lines connect two end nodes (A and B). The billing portions of a leased line include:

access from location A to the carrier's point-of-presence (POP)

a mileage sensitive charge from the carrier's POP serving location A to the carrier's POP serving location B

access from the POP serving location B to location B premises.

A leased line can also be served from a company's T1 network, but the circuit still requires bandwidth.

A leased line can run at 100% of its capacity with no billing ramifications, although performance will suffer. We will monitor leased line performance to trend average utilization.

We will look for the following thresholds:

Averaging less than 20% utilization. This may mean that circuit speed can be reduced if the circuit is running at Fractional T1 speeds. We will determine potential cost-savings and present them to you.

Averaging 20% to 60% utilization. This is a good utilization level, no action will be required.

Averaging more than 60%, with spikes to 85%. This means that the circuit speed may need to be increased if application requirements increase. We will determine costs associated with increasing circuit speed, and present those costs to you for budget consideration.

Averaging more than 80%, with spikes to 90-95%. This means that circuit speeds should be increased promptly. We will determine costs and present those costs to you.

If utilization is averaging more than 90%, with spikes nearing 100%, you have probably already heard numerous user complaints. This should be an immediate action item.


**Frame Relay**
Frame Relay billing is comprised of:
Access (or circuit) - this is mileage and speed sensitive
Port Speed          - you are billed on port speed increments
CIR                 - increasing CIR increases costs
PVC                 - you are billed for each PVC

We will monitor frame relay utilization using vendor provided tool sets. Major frame relay vendors have utilization reports available. We will request utilization reports from your frame relay vendor and look for the following items:

Serial port utilization
Dropped DE bits
Dropped Non-DE bits
PVC throughput
PVC throughput percentages

As you are billed for port speed, we will monitor utilization. A low utilization level (averaging less than 20%) can point to an opportunity to decrease the port speed. A high utilization level can point to a need for increased bandwidth.

By looking for a high number of dropped bits we can determine if CIR is too low, and needs to be increased. If no bits are being dropped, there is a strong possibility that the CIR is too high, and can be reduced.

We will look at PVC throughput. This will show how much PVCs are being used. By looking for highly utilized PVCs, we may be able to determine if additional PVCs should be built for specific applications. Alternatively, if the network is fully meshed and PVC utilization is very low, we may have an opportunity to remove PVCs.

Vendor reports will be supplemented with a network monitoring tool designed for WAN trending. This will give us greater report detail, and a more granular reporting level. Some vendors report on 4 hour averages. Some vendors report on average utilization during the busy hour. By using our own tool set, we will be able to determine minute to minute requirements.

### ISDN
ISDN offers a high level of flexibility in meeting networking requirements. By using ISDN, a business can change connectivity on a minute-by-minute basis. Unfortunately, every one of those minutes is billed to you.

We will monitor ISDN usage to determine the average number of minutes per day, per week and per month that each remote office or telecommuter is connected.

We will determine the costs by remote office or telecommuter. This will include the fixed ISDN line charge, and usage sensitive charges.

We will compare ISDN costs for high-usage offices with frame relay costs, to determine the most cost-effective way to provide connectivity to high volume users. We will include new equipment costs to present a break-even point, then determine cost savings after the initial payback period.

**Modem Pools**

We will verify how modem users are connected currently. If modem users are connected directly to business lines, we will determine if there are available analog ports on the company phone system. If not, we will determine the cost to provide additional analog ports, and determine the break-even point for converting modem users off business lines and onto the company phone system.

We will monitor dial-in usage to determine the appropriate number of analog lines to serve modem pool users. This can be outbound calls, or inbound calls to an e-mail server or network application server.

We will verify how modem calls are placed (long distance DDD, calling card, ...) and determine the most cost effective way of providing a dial-in solution.

## Network Optimization - Phase I

After completing all network auditing, documentation, and measurement tasks, *Elizabeth-Bruce Communications Consultants* will begin a network optimization phase.

Each phase of the network optimization project is optional, although Phase I is a pre-requisite for Phase II. We can complete Phase I for you, and provide you with documentation for self-managing Phase II, or we can perform both phases.

The first part of the network optimization plan includes creating a **Request For Information** (RFI). This requests general information from major network providers, and will give you a background on the major carriers.

We will also create a **Request For Quotation** (RFQ). This is a simple request for pricing for specified services at specified locations. Responses to this document will give you an estimate of prices for performing a network optimization. At this point we will be able to decide if the estimated cost-savings associated with the network optimization are worth the expenses in equipment, time and installation fees.

If the replies to the RFQ look promising, we will create a detailed **Request For Proposal** (RFP). This will be sent to the most promising carriers. The RFP will detail your network requirements. The RFP will also detail your reporting requirements and service level requirements.

Replies to the Request For Proposal will be carefully reviewed, and advantages and disadvantages of each solution will be documented. We will assist in vendor selection, but as working relationships are sometimes more important than technological solutions - we will work closely with representatives from your user community and your networking staff. We will facilitate, we will not direct.

After the network staff and representatives from the user community reach agreement on which carrier to use, we will create a project plan. The project plan will address:

> Circuit requests
> Monthly circuit costs
> Circuit installation costs
> Projected circuit installation dates

Dmarc responsibilities
Projected costs to extend Dmarc to computer rooms

CSU/DSU requirements
CSU/DSU costs

Router requirements
      Configure spare serial port, or perform a hot-cut
      Estimated down-time associated with hot-cut
Cable requirements
      Ship spare cable, or perform a hot-cut
      Cost of spare cable vs. Down-time

Projected cutover date
Cutover instructions for remote offices

Circuit disconnects - old network

Contacts:

      Site contacts
      Network carrier contacts
      Help desk contacts
      Network staff contacts

Responsibilities
      We will document who is responsible for each item


*Elizabeth-Bruce Communications Consultants* will create a complete project plan. The project plan will be used for Phase II of the network optimization project. We will perform Phase II of the network optimization project if you choose. Alternatively, if you wish to keep your costs low, your staff will be able to use the project plan and perform phase II in-house.

## Network Optimization - Phase II

Phase II of the network optimization project consists of the tasks required to migrate the network over to a new carrier. Following is a brief outline of tasks. Information from Phase I will be used to complete Phase II requirements.

## Pre-Cutover

Before converting the network, we will take the following steps:

### Router Related:

Telnet to remote router

> Verify servers and SAPs router currently sees
> Verify routes router currently sees
> Verify input and output IP filters
> Verify IPX network filters
> Verify IPX SAP filters

By telneting to the router and verifying what the router currently sees, we will be able to make sure that no users are affected by the cutover. We will verify all servers that the user community can see, and can log on to.

We will document all information above, and compare results after the cutover to make sure filters are applied appropriately.

### Carrier Related:

Place circuit requests with carrier
Verify Dmarc is to be extended to computer room by Telco
Verify CSU/DSU provider and responsibilities
Verify CSU/DSU to router cable interface
Coordinate CSU/DSU installation when circuit is in
Verify circuit tested clean end-to-end

By working closely with the carrier, we will be able to keep track of circuit requests. We will also be able to make sure that installation of the CSU is not held up, as we will make sure that the Dmarc is extended to the computer room. Normally *Elizabeth-Bruce Communications Consultants* uses the Telco to extend the Dmarc. That way the Telco is responsible for the circuit all the way to the computer room. We also normally use a CSU/DSU provided by the network carrier. That way the provider is responsible all the way to the CSU, and limits problem determination during network outages.

An alternative would be to use a cabling company to extend the Dmarc, and to purchase and configure CSUs. While it may be more expensive to have the Telco extend the Dmarc, and it may be more expensive to have the carrier provide the CSU - it is worth it the first time the network goes down.


**Coordination:**
Place configuration requests with in-house software group
        OR subcontract
Notify site contacts
Place circuit disconnects (with push-out dates)
Ship spare cables to each site that has router with spare WAN port
        Include instructions for connecting cable
Keep in contact with network provider on Telco dates


Coordination activities help to make the cutover less eventful. We will work with the network staff to make sure they are aware of all changes that will hapen, and when they will happen.

We will also place requests to modify router configurations to support the new network. This will include DLCIs, circuit speeds, and PVC mapping. If you do not have a group responsible for network configuration, we will work with a subcontractor.

We will keep site contacts informed as we make progress. We will notify site contacts of circuits that will be installed, and projected installation dates. We will also notify site contacts of where the Telco should install the circuits.

Circuit disconnects normally take 30 days to complete. Sometimes they take longer. We will place circuit disconnects 10 days before new circuits are due to be installed. We will request circuits to be disconnected 30 days after our requests are made.

This will give us 20 days to make sure new circuits are in, Dmarcs are extended, CSUs are installed and configured, circuits are tested, and routers are connected. If it looks like we will miss the projected online date, we will push out the circuit disconnects.

In offices that have routers with a spare serial port, we recommend purchasing a second serial cable and shipping it out. We will coordinate with the site contact to connect the cable to the router, label it, and leave it hanging. We will coordinate with the network vendor to connect his CSU to the cable. The cost of a serial cable is about $100. This will let us avoid any down-time at all.

An alternative is to perform a hot-cut. The office will have no network connectivity while they are converting from one carrier to another. The average down-time is 20 to 30 minutes, although outages can be much longer if the carrier has any network problems.

Telcos vary greatly with service, or lack thereof. We will keep in contact with the selected network provider to make sure that the Telco will meet projected dates. In areas where it looks like the Telco will miss a date, we will work to escalate the issue as it becomes more critical.

**Cutover:**
>Coordinate CSU/DSU connection to router with site contact
>Coordinate port activation with carrier
>Verify circuit up, protocol up, passing data
>Verify IP routing
>Verify IPX routing, servers and SAPS

Through careful pre-cutover planning and coordination, the cutover should be uneventful. We will call the network vendor to make sure the circuit is installed and tested. We will also make sure that the CSU is installed and configured.

In sites that have routers with a spare serial port, we will telnet to the router and make sure that the port is configured for the proper protocols. We will also verify filters applied to the interface.

In sites that have spare serial ports, we will talk the site contact through verifying that the spare cable is connected to the proper CSU and to the proper serial port on the router. These offices will not experience any down-time related to the network conversion.

In offices that do not have spare serial ports, we will verify that a software specialist is available to reconfigure the serial port before we move it to the new carrier. We will check and make sure that the modem is attached to the router and that it accepts dial-in access. We will take the following steps to cut over an office that does not have a spare WAN port:

Contact network vendor.
Make sure circuit active, port active, DLCI mapped

Contact network specialist
Make sure ready to dial in to modem
Make sure ready to reconfigure serial port

Conference site contact
Let contact know will have 20 - 30 minute outage
Talk through moving cable
Have network specialist load new configuration

Verify port comes active
Verify IP routing
Verify IPX routing
Verify filters

## Post Cutover
        Update documentation systems
        Create new network map in Web server format
        Monitor utilization, througput, dropped packets
        Monitor service level (outages, response to outages)
        Verify disconnects placed
        Verify new network bills reflect proposal prices


After the cutover, we will make sure that the Information Management system is updated with new circuit IDs and port connectivity.

As the network changes, we will create new network maps in Web server format, to be posted on your company's Web server. This will help keep the network staff informed as the project progresses.

We will also monitor utilization, to make sure the new carrier is delivering services as proposed. We will monitor utilization statistics, and throughput statistics.

As network outages are an unfortunate part of life, we will monitor the network carrier's response to outages. We will monitor the number of outages, the severity of outages, and the mean-time-to-repair. If this deviates from the service level agreement, we will seek financial remedies.

We will verify that all circuit disconnect requests have been placed, and have been processed. We will also verify network bills to make sure that old circuits are no longer being billed to you.

We will review all new network bills to make sure the new vendor's billing agrees with what was proposed.

# Cost Justification

While a thorough network audit can reduce your network bills, it can also improve network performance. *Elizabeth-Bruce Communications Consultants* specialize in the fields of:

Network documentation
Network performance tuning
Network utilization tuning
Network optimization

We feel that each activity we perform can stand in its own right.

A network audit can reduce your bills.
Network optimization can improve performance and reduce bills
Network documentation can increase network availability

Billing will be based on the successful completion of each module, using the percentages noted below. *Elizabeth-Bruce Communications Consultants* will bill on an hourly rate, with a not-to-exceed figure. We will determine the not-to-exceed figure after performing an initial overview of your network.

Network Audit
Completion of circuit inventory - 10%
Completion of network equipment inventory - 10%
Update information systems - 10%

Network Optimization
Completion of network utilization reports - 10%
Completion of circuit disconnects - 10%
Completion of circuit speed upgrades/downgrades - 10%

Network Migration
Completion of Phase I     - RFI, RFQ, RFP      - 20%
Completion of Phase II    - Network conversion - 20%

If site surveys are required, *Elizabeth-Bruce Communications Consultants* will follow your travel and living guidelines. All normal travel and living costs are to be reimbursed *to Elizabeth-Bruce Communications*.

Any carrier associated charges for reports are to be paid by the client. All such reports will be given to the client after the network audit is completed.

# Schedules

*Elizabeth-Bruce Communications Consultants* will begin the network audit within 30 days of acceptance of this proposal. As the length of time required to perform a network audit is dependent on the complexity of your network and the complexity of your Information Management systems, we will establish a timeline after performing a preliminary overview of the network.

We will keep you informed of progress by issuing weekly status reports.

The following items are pre-requisites to the successful network audit:
    List of office addresses
    List of office contacts and phone numbers

    Assigned representative from your network staff
    Assigned representative from accounts payable staff

    Browse only password to routers
    Browse only password to Netview
    Userid and password for your information management system
    Access to any network management workstation you may have

    Network carrier contact

# Further Reading

Creating RFPs
Mikulski, Managing Your Vendors, PTR Prentice Hall

Proposals
Sant, Persuasive Business Proposals, AMACOM

Voice Auditing*
Brosnan & Messina, The Complete Guide To Local & Long Distance
Telephone Company Billing, Telecom Library Inc.

T1 Networking*
Flanagan, The Guide to T-1 Networking, Telecom Library Inc.

Frame Relay*
Heckart, The Guide to Frame Relay Networking, Flatiron Publishing Inc.

ISDN*
Flanagan, ISDN A Practical Guide To Getting Up and Running, Flatiron
Publishing Inc.

Newton's Telecom Dictionary*
Newton, published by Flatiron Publishing Inc.

---

* All books are available for direct sale by

Flatiron Publishing Inc.
12 West 21 Street
New York NY 10010
212-691-8215 or 800-LIBRARY or www.telecombooks.com